James Frederick Sarg

A new dairy industry

preparation and sale of artificial mothers' milk

James Frederick Sarg

A new dairy industry
preparation and sale of artificial mothers' milk

ISBN/EAN: 9783337414283

Printed in Europe, USA, Canada, Australia, Japan

Cover: Foto ©berggeist007 / pixelio.de

More available books at **www.hansebooks.com**

A NEW DAIRY INDUSTRY

Preparation and Sale of Artificial Mothers' Milk

"NORMAL INFANTS' MILK"

BY JAMES FRED. SARG

Late of Hessenhof, Lake Constance, Germany

BLACK FOREST FARM

KEMPSVILLE, VA.

U. S. A.

NORFOLK :
W. T. BARRON & CO., PRINTERS
1896

CONTENTS.

INDEX.

INTRODUCTION.

IT has always been the investigations of science that
have graded the path on which practice has followed, but too often sluggishly after a longer of
shorter time; it has been the same in regard to the
production of a rational nourishment for infants.
Here science has recorded singular successes on the
different fields that must contribute to the attainment
of a desirable product, but practical execution has
been slow to follow the lead.

Statistics have forced upon us the conviction that
the mortality of infants artificially nourished is so
much greater than that of those nourished in the
natural way—on the breast, and that whatever difference there may exist in the causes of deterioration
in the various levels of human society women live in
amongst civilized nations, the fact is uniformly established that the development of the milk glands in
the female breast is steadily decreasing.

Cow's milk will, for general purposes, ever be regarded as the best substitute for mother's milk. Natural science has done much to impart the knowledge
of the influence of feed on the production of milk,
and engineering has, by the invention of improved
machinery, perfectly revolutionized dairy technics,
while the production of a healthy infants' milk has

encountered its greatest difficulty in the conservatism of the farmer, who is slow to adopt advice or change his methods.

The production of normal infants' milk is a field of work that stretches over so many industries and sciences that a thorough mastering of them can impossibly be expected of the dairyman who would undertake the manufacture of " normal infants' food," but a familiarity with the scientific principles of all and every operation comprised in the manufacture should most decidedly form a fundamental part of his stock in trade. Referring to this sentiment, I will beg my readers kindly to bear in mind that I am a farmer writing for farmers.

I have to thank Dr. H. Weigmann, of Kiel, for the permission kindly granted to translate from his excellent work the bacteriological part of this treatise, which I herewith recommend to the indulgence of all those who are, and also of those who should be, interested in the amelioration of the conditions for producing a healthy food for infants.

JAMES FRED. SARG.

Black Forest Farm, Va.,
October of 1896.

CHAPTER I.

Milk and Milking.

Those organs whose secretions we give the name of milk are called milk glands and their aggregate form in the cow, including the skin that covers them, the udder.

These glands do not, by nature, come into activity until a short time before parturition and during a variously protracted period after this act. The first secretion in the udder caused by a heightened afflu- ence of blood to all generative organs after conception, is noticeable about the middle of the period of gesta- tion; the teats of the heifer will at this time, when stripped, render a small drop of viscuous transparent gum, which when ocurring may be accepted as the first visible sign of pregnancy. This sign does, how- ever, not repeat in the cow. Differing from other animal secretions milk is opaque and, when healthy, of a white color. Other hues of color with exception of the first or colostral milk, which is of a yellowish tint, indicate rapidly decomposing milk or the pres- ence of bacteria; some few intensely colored vegetable foods are also able to give a coloring to the milk. The agreeable sweetish taste of normal milk may be changed by the influence of food or by diseases of the udder. An inflammation ascribed to the action of a bacterium of the streptococcus species produces a salty taste in the milk which at such time is also slimy.

Bitter milk is not infrequently noticed in cows with a protracted lactation—but may be an effect of food given; it has been noticed, for instance, after feeding large quantities of young clover and always indicates the presence of micro-organisms.

The smell of freshly drawn milk is faintly like that of the skin of the animal and is probably produced by the presence of etheric acids of fat.

The reaction of milk is generally "amphotere," which means to say that it will turn blue litmus paper red and also turn red litmus paper blue, a condition based on the simultaneous presence of neutral and also of acid alkaline phosphates and calcium caseinates; one of these predominating turns the reaction to that side. Boiled milk acquires an intensified alkaline reaction. The boiling point of milk is about 1° F. higher than that of water, and its freezing point is 1° below that of water.

The specific gravity of milk, dependant on its temperature, varies with the relative quantities of its composing elements: water, butterfat and solids. Instruments have been invented to ascertain the specific gravity, for

Plain Lacto-densimeter.

instance, the lactodensimeter of Guevenne and Soxhlet. By the aid of the specific gravity, with a known amount of fat, the solids may be calculated. These

instruments are valuable as a means to detect watered
or skimmed milk. The specific weight of milk
ranges from 1.027 to 1.035. Colostral milk at 60°
F. 1.056; skim milk, 1.032 to 1.037; cream,
on an average, 1.010.

Amongst the chemical ingredients of milk
we find all the principles of nourishment:
proteids, fats, carbohydrates, salts and water.
Amongst the albuminoids in the milk casein
predominates. It is accepted as probable by
some that the casein in cow's milk is
identical with that in human milk, although
we note that the casein in woman's milk,
when coagulated by the action of rennet,
is by far more fine-flaked and jellyfied than
that from cow's milk, which latter forms
into compact solid flakes. The difference of
coagulating is probably due to the different
quantity in which salts are present in the two
milks; but this distinctive difference in coagulating, we must bear in mind, constitutes one
of the principal deficiencies when we come to
look at cow's milk as a substitute for mother's
milk. This is of such salient importance in
the transformation of cow's milk into artificial
mother's milk, that the closest study of the
various investigations carried on at the present

Lactodens-
imeter with
Thermome-
ter.

time on this line must be recommended to all that
would undertake the manufacture of normal infants'
milk. Cow's milk and human milk differ with re-

spect to the curdling of the casein, the content of
salts, the absolute content of nutrients and the relation of the various constituents. The nature of the
coagulated casein in the stomach depends upon the
casein solution, the content of soluble calcium salts
and the acidity of the solution. Cow's milk is in
these three respects unfavorable to the best coagulation, for it contains twice as much casein, six times as
much lime and is three times as acid as human milk,
while this latter contains but one-third as much of
acid phosphates as cow's milk.

Casein forms three chemical compounds with calcium or sodium—dependent on the predominant reaction—the mono, di and tri-calcic (or sodic) casein.
Only the dicalcic or disodic casein compounds are
curdled by rennet in the presence of water soluble
lime salts, and the completeness of the curdling depends on the amount of lime salts ; we may, therefore, attribute the compactness of the casein curdling
in cow's milk to an increased alkalinity. The studies
of *Bechamp* show that casein is not a soluble substance which may be coagulated by acids, but that it
is an insoluble substance forming soluble compounds,
caseinates, with alkalies and lime, and that the insoluble casein may be precipitated from these compounds by acids which combine with the bases of
caseinates. The change in the casein by the action
of rennet has no connection with the reaction. We
shall see later what effect heating produces on the digestability of casein and on the milk proteids in general.

Further albuminoids of milk, but of secondary importance, are lactoglobulin, lactalbumen and peptone, the nutritive value of which is, however, considerably impaired by boiling the milk, by which a greater part is changed to hemialbuminose.

Following the albuminoids, the different fats in milk merit our attention; we designate them collectively as butter-fats, and find them suspended in the milk in emulsive condition, that is, globules of the minutest size; these globules, coated with casein, give the white color to milk. The size and number of globules is variable in one and the same animal, being affected by the advance of lactation, change of feed and by sickness. With the advance of lactation, the number of large globules diminishes and that of the small globules increases; with the change from dry feed to green feed in the spring, there is an increase in the proportion and the number of the large globules. Disease or sickness and the use of cows for draft, when not accustomed to it, has a marked effect in diminishing the number and size of globules. Succulent food decreases the size and increases the number of globules; oats, bran and linseed meal increase their size. Age is apparently without effect. Morning's milk has larger globules than evening's milk. The first part of the milking has fewer and smaller globules than the last.

Butter-fat is liquid at from 85° to 105° F., when cooled below 60° it becomes of a crumbly consistency; notwithstanding milk may be cooled to 32° without the suspended fats becoming hard, only below 32^c or by mechanical agitation the form of the globules is lost, they become solid and their contour rugged.

On standing, the globules rise to the surface by virtue of their minor specific weight and they form the cream, while the milk beneath it is termed skim milk, which, however, is not entirely free of fat, because the minutest of the fat globules find it impossible to push through the viscuous milkfluid to reach the top. Warmth favors the ascending of the globules, cold retards it, but we avoid the warmth because it involves a rapid decomposition of the milk. A large number of instruments have been invented for the purpose of ascertaining the quantity of fats; some of them aim to accurately measure the quantity of cream raised in twenty-four hours and are called cremometers, others purport to ascertain the percentage of fats by diluting milk with water and making it translucent until a certain mark on the instrument is visible; these are termed lactoscopes. By far more exact and scientifically correct is the method of *Soxhlet*, who ascertains the specific weight of fat in the milk; his apparatus is, however, too complicated to be of much use outside of the chemist's laboratory.

Cremometer.

The method that gives the best results for practical working of the dairy industry is the one that dissolves the casein by an excess of acid under the influence of heat and rotatory motion. The best known in this country of this class is the Babcock tester, the use of which I shall describe further on.

TEST SET FOR CREAM.

We need not go into the detail of the different fats or fat acids composing the butter-fat, such as butt-yrine, capronine, capryl, laurine, myristine, palmitine, stearine, arachine, olein and glycerine acid, further than to remember that it is the varying amount of these fat acids contained in the feed we give the cow that produce the varying degree of either firmness or grease-like consistency in the butter.

The color of butter also is largely dependent on the relative predominating of one or more of the above-named fatty acids.

Another characteristic ingredient of milk is milk sugar. Under the influence of different ferments, among which principally the baccillus acidi lactici is noted, milk sugar is transformed into milk acid.

Milk sugar is sometimes attacked by a rosary-formed species of a coccus, engendering a slimy fermentation, which results in what we know as slimy or long milk, which is generally unfit for the extraction of butter, because the minute fat globules are unable to rise in this viscuous fluid and form the cream.

In connection with these ferments, it may be mentioned that some of them, like the Sacharomyces cerevesiæ and the Dispora caucasica, are used to bring milk to an alcoholic fermentation, in which state it possesses intoxicating properties, and by reason of these is valued as a beverage and largely consumed by various tribes of Turkestan and Circassia under the name of Kumys and Kefyr.

Other organic matter contained in milk is a minute quantity of citric acid, a number of aromatics, like anisol, cuminol, cymol, tymol, in fact, all such as are found in the food of herbivorous animals and traces of fibrin.

Of anorganic or mineral matter, it is principally sodium and phosphoric acid that merit attention, as we know that cows with protracted periods of lactation are deficient in these ingredients. When we, therefore, consider that a healthy and normal formation of bone in a child is in great manner dependent on the unstinted assimilation of phosphoric acid in its milk, we see the justice of refusing the milk of such animals whenever the manufacture of infants' milk is aimed at.

Quantity and quality of milk are, as we may suppose, greatly influenced by the quality of food, the management of the feeding and the breed and individuality of the animal.

Medicinal qualities contained in the food or pasture eaten by the cows may reappear in the milk and trouble the consumer; for instance, the feeding of cabbage leaves to cows produces flatulence and pains in most infants which consume such milk; also the acidity of feed like that in wet and acid brewers' grains passes into the milk and makes it unfit for infants' food. Increased feeding of albuminoids favors an increased production of fat in the milk, while a feeding with a preponderance of carbohydrates is followed by a loss of albumen and fat in the milk. The quantity of milk is influenced also by the periods of lactation; immediately after parturition it is at its height, and from that time decreases generally, not gradually but in about three well defined periods the duration of which is naturally dependent on the entire duration of lactation, which,

Dairy Thermometer.

as we all know, is exceedingly variable, both as to every separate animal as also in the several lactations of one and the same animal. A lengthened period of lactation is acquired by heredity and confirmed by judicious management at the hands of the milker.

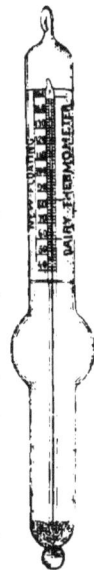

Concerning the qualitative changes of milk during the period of lactation, there is no harmony of opinion prevailing, yet a majority of investigators claim

that towards the end of the lactation the percentage
of solids and of fats grows. With reference to the
time of day at which it is drawn, it is generally conceded that in barn feeding the quantity of morning's
milk is larger than that of the evening's milk, but
that the latter is richer.

Spaying, the removal of the cow's ovaries by a sur-
gical operation, has the effect to prolong the period
of lactation, in some instances which are on record
for a time of three years running, and upward. The
length of the period of lactation is one of the most
important factors in judging the value of a cow, but
for obvious reasons castration should only be executed
on such animals as by nature are arriving at the close
of their remunerative career or of their generative
functions.

From the foregoing we should receive the impression that the udder of the cow is a valuable machine,
one whose handling should be thoroughly understood
by every person—male or female—called upon to
work it. Where is the wisdom of spending a large
sum of money on a superior cow if her udder is to be
handled by an ignorant and careless milker? In
every other trade we expect from the workman,
and even from the apprentice, an exact knowledge
and familiarity with the tools he uses and with the
processes embraced in the application of his trade.
The average farmer or dairyman, however, seems to
be an exception to this rule, if we may judge by the
lack of knowledge he possesses as to the physical

make-up of the cow. Drawing the milk from a cow
seems an operation of such absolute simplicity to the
mind of many that nothing can be said about it more
than they already know, and yet an ignorant milker
is apt to spoil the best cow in a short time.

Milking is generally done on the right side of the
cow. The milker sits on a low stool which in different localities has one, two, three or four legs, the
milk pail pressed and held firmly between his knees,
his head inclined against the paunch of the cow. The
cow's tail may be secured by some device and pre-
vented from striking the milker's head, but unless
flies are very bad it should be left loose. The milker's
hands should be scrupulously clean. Whether the
milker's hands should be wet or dry is an open question, as both methods are quite extensively practiced.
Milking with a dry and dirty hand is, perhaps, a
cleanlier operation than milking with a wet and dirty
hand. We have the painful conviction that a greater
number of cows are milked with dirty hands than
with clean hands and it may be, therefore, safer to
advocate the use of the dry hand.
However, when milk is drawn with
intention to manufacture it into infants' food, and the necessary precautionary measures for cleanliness are
strictly observed, milking with the wet
hand (that is to say, putting a few drops
of milk in each hand) may be adopted with considerable advantage to the animal, because the operation

Milk Pail and Strainer.

is then not so irritating to the subcutaneous nerves of the teat and udder. Then, too, a sore and bruised teat may by the wet hand be milked without pain to the cow, while the dry hand may produce restlessness. Lastly, it may be claimed that the wet hand comes closer in imitating the function which nature expected the teat to be used for—the sucking by the calf's mouth.

A method which finds its place between the two just mentioned, and which is extensively practiced in Switzerland and Southern Germany, is to milk with the dry hand, but to apply a small quantity of pure lard about the size of a large pea—to the fingers and thumb—the application to be repeated with each cow milked. The lard is carried around in a small metal cup fastened to or around the leg of the milk-stool.

The milker should grasp one front teat and one back teat of opposite sides of the udder so that the emptying of the two halves of the udder proceed simultaneously. Owing to the position of the milker's head, the milking cannot be followed with the eyes, therefore he must be guided by the touch and hearing ; for this reason all loud conversation or other vociferation should be interdicted during milking time, because this gives occasion to interrupt the milking. Apart from the loss of time, the interruptions are not good for the cow because they multiply the nervous irritation, causing the animal to become restless, which should be avoided. Many of the best milkers are accustomed to hum a tune while milking, and this is an excellent practice, as it has a plainly apparent soothing effect on the cow.

To learn to milk well it should be practiced slowly, because both hands must become equally expert; the pressure of the hand on the teat must be applied in regular alternation, so that when one hand closes around the teat the other hand opens, and the flow of milk into the pail is continuous; an experienced ear can detect at once if a milker works well.

The full hand should grasp the teat as high up towards the udder as possible, then the thumb and index close tightly around the teat so as to shut off the milk contained in the teat from retreating into the milk cistern when the pressure on the teat is applied. Then the other fingers, one by one from the index downward, close around the teat in rapid succession and press out the milk. The amount of pressure required to press the teat depends on the more or less developed muscles that encircle the orifice of the teat for the purpose of retaining the milk, which would, without this provision, flow to the ground as fast as produced. Cows in which these muscles are strongly developed are called hard milkers. As soon as the milk has been pressed from the teat, the hand eases up, and immediately the milk from the cistern rushes into the teat, filling it again; the pressure of the hand and fingers is repeated until the firstly grasped pair of teats do no longer give a full flow, whereupon both hands change to the two remaining teats. During the rest now given to the first milked pair of teats, the milk has time to collect from the remotest cells of the glands and fill the milk cistern anew. This changing of hands to alternate pairs of teats is repeated as long as milk will come, and should be continued without interruption. The more rapid and the more symetrical the work can be performed, the better the cow will allow herself to be milked, the more and the richer milk she will give. The upward motion of the hand at every repeated closing round the teat produces a kneading motion on the udder, which is of great importance to keep the milk in the cistern in commotion. When the flow of milk seems to have been exhausted by the milking, then each teat is taken between the thumb and index finger and "stripped" downward. This should be done merely to insure an absolutely thorough removal of all milk from the udder, and should never be resorted to when the udder is filled, because it is apt to spoil the udder. Careless removing of all milk from the udder will result in serious damage, because it has, aside from the loss of the milk, a deleterious influence on the glands, tending to interrupt the productive action in the minute cells where the milk is formed. An extended period of lactation has been bred into cows, and we should try to confirm this habit by milking the heifer after her first calf as long as possible, even if the quantity of milk given is, in time, only a small one, because, allowing her to dry off too soon before her second calf, this habit of drying up is soon confirmed.

Milking is a tiring task and not too many cows

should be apportioned to the milker, because a tired milker does not do good work, particularly as some cows are difficult to milk ; some have an uncommonly small orifice in the teat, some have strong closing muscles ; others, again, strive to retain the milk entirely. This may happen in consequence of the cow feeling pain from the milking as, for instance, in sore teats, or she may be afraid of ill treatment, or try to retain the milk for her calf. To find an explanation for this voluntary retention of the milk we must go into the anatomy of the udder. We have already mentioned the muscles closing the orifice of the teat, we shall now see that a large quantity of blood is brought from the heart to the udder in strong arteries, which, branching out into the minutest vessels, spread through the entire milk glands, enveloping the minutest cells and engendering their action of producing milk, and that this blood is led back again to the heart by an equally complicated system of veins that are spread over the entire inner surface of the udder, even down to the point of the teat enveloping the entire tube or duct of the teat with a network of veins. If the cow now retains her breath she produces a check on the flow of blood which tries to return to the heart, and, in consequence, the veins in the udder become swollen and therefore help to close the orifice and duct ; if she manages to repeat this retention of breath—in short repetitions—she is able to suspend the flow of milk entirely. The remedy for this bad habit is either to give some mash or

3

drink which the cow likes, or to fasten a bunch of straw in her mouth; or, what is nearly always the most effective, to treat her with quietness and patience, at the same time milking persistently. If another person is present to stroke along the under part of the cow's neck she will give up the retention of breath at once. When a cow retains her milk on account of pain as, for instance, with chapped teats which frequently occurs during first spring pasture, a remedy is only found by kind treatment and milking rapidly with a soft hand. Such teats should be carefully dried after each milking and an ointment applied. Whenever the milker has any reason to suspect any derangement of the cow he should taste the milk from every teat and look at its color; any carelessness in this respect may result in spoiling the milk from the whole stable. As a rule, milking should be performed only morning and evening, making the intervening time as equal as possible.

As to the advisibility of feeding during milking time there are many reasons against its being adopted. When cows are once used to being milked before feeding they are much quieter and the business is concluded much more rapidly; but there are other reasons of importance, as we shall see later, for not feeding during milking time, particularly for not giving any dry roughage.

The dexterous strong hand will always be the best milking machine; only in case of disease the milking tube should be made use of and no other milking

machine of any kind should be applied. One of the
most essential requisites during the times of rest for
the milk cow is absolute quiet, guarding her against
fright and preventing worrying or violent exertion.
A great deal has been said and written about the necessity of giving cows daily exercise in the open air, and
though nothing is to be said against pasturing in fine
weather, it is certain that in very hot or in cold and wet
weather the stable or barn is the only proper place to
keep the cow in. Every exertion, therefore also that ne-
cessarily combined with locomotion, is an expenditure
of force, a wear on the muscle, and this wear must be
replenished by an extra amount of feed, the quantity of
which will be found in exact relation with the distance that has to be traveled over and the time consumed by the animal until it has been able to graze a
sufficiency for its needs. It is easy to see that a cow
which is enabled to eat all she requires in one hour's
time and can then lie down, in perfect rest, to ru-
minate and digest, is in an eminently better position
to turn her food into milk than the cow that has to
walk about, for three or four hours at a time, grazing
before the feeling of hunger leaves her. Nothing
should, however, be more strongly condemned than
the practice of leaving cows in the open air during
midday in hot Summer weather. Not only does the
intense heat of the sun tend to harden the skin, con-
tracting the pores, and thereby diminishing the general vitality of the animal, but also the constant
irritation produced by flies and like insects has a

notable and injurious effect on the milk production, which will be the more easily noticed the higher the nervous system of the cow, as an individual, or as a member of her breed, is strung. Also the sexual functions are often seriously affected, postponed or obliterated by this irritation.

Having now acquired a cursory idea of how milk is formed, and how it should be drawn, let us turn to the influences which tend to spoil it, the methods employed to counteract these influences and give milk good keeping qualities.

CHAPTER II.

The Origin of Bacteria in Milk and the Conditions Favorable for their Breeding and Multiplying.

It is a well known fact that milk undergoes a radical chemical change only a few hours after it has been drawn. This change, to our visible conception, consists in the milk becoming sour, in other words the milk sugar has changed to milk acid and, in consequence of this acidity, the casein has been separated from its connection with lime and is set free—the milk "curdles." We generally notice only this first phase, because in itself it is sufficient to unfit milk for further use. A second phase follows in which the casein is partly dissolved and fermentation sets in, bubbles of gas forming, and the process is wound up with real putrid decomposition and the forming of mould.

In the microscope we possess an instrument that enables us to enter into a study of the composition and life of the lowest organisms, and also a means to enable us to make and study their culture, through which it has been demonstrated that every process of decomposition of organic matter is due to the action of such organisms and that they, somehow, disin-

tegrate the more complicated matter and are able to
reduce it to the primary ingredients of composition.

When we look at a fluid or other matter in a state
of decomposition, under the microscope, we notice
strewn over the entire field a complexity of threads,
longer and shorter tubes or cylinders and egg-shaped
bodies, and going on between all these is seen a slug-
gish rotatory movement of one or more of the chain
of cylinders, possibly, too, a worm-like movement of
the spiral threads. By the means of different cultures
we are able to separate the several organisms of
this intricacy, when we shall find that the spiral
threads and the small tubes are parts or spores of a
mould fungus, and the small oval bodies are probably
ferments, while those that we saw in the most active
motion belong to a series of organisms which have
one peculiarity in common—they multiply with ex-
traordinary rapidity by breaking up into pieces and
every one of these pieces forms a young germ. Every
liquid, be it of animal or vegetable origin, when exposed to the air, contains a large number of such
organisms. Milk is no exception and it contains
them not only when it commences to turn to visible
decomposition but immediately after leaving the
udder, yes, even in the lower part of the udder itself.
Thus it is easily explained why milk decomposes so
rapidly after having been drawn. How and by what
route do these organisms enter milk? Are they
already present in the glands of the udder or do they
enter the milk later? These questions can be posi-

tively answered by the assertion that the glands of a *healthy* cow give off milk absolutely free from such organisms. We call such milk sterile. Germs enter, manifestly, from the outside and may therefore be termed a pollution of the milk. These decomposing germs are encountered in great abundance where organic matter is in the act of disintegrating into its composing elements, and of such decomposing matter there is enough around the premises where we draw milk—the stable ; there is, in fact, generally more than necessary, and this is easily brought into contact with the outer cover of the milk glands—the udder. The location of the udder of our domestic animals involves a continual exposure to its being soiled by the excrements, urine, dust from the bedding, and even our most scrupulous cleanliness and precaution cannot prevent, during milking, a quantity of dirt, particles of straw and fodder, dust, hair and excoriations from finding their way into the milk. It may, therefore, be taken for granted that the greater part of dirt, and, therefore, the greatest mass of spores, is derived from the udder, as well from the external part of it as from the openings in the teats, and even from the interior milk cisterns. Dairymen know well that the first strippings when commencing to milk are by no means favorable for the making of cheese, and in many dairies I have found it customary to milk the first few strippings into the bedding. Many of the germs possess very active motion and from a soiled teat find their way into the interior of the duct. Investigation has proven that the first milk drawn contains about fifty to eighty thousand bacteria to a tenth of a cubic inch, while the next following or, we may say, the bulk of the milking contains about five thousand to the same quantity, and only the last quarts drawn are nearly or entirely free from germs. An immigration of germs by way of the teats cannot be doubted and is the cause, not infrequently, of some forms of inflammation of the udder.

As we have seen, milk is already polluted at its exit from the soiled udder, and again by the dropping in of dirt from the external part of the udder, and when we consider that dung is nothing more or less than the undigested residue of the fodder eaten, filled with unutterable numbers of bacteria and spores, we are then able to draw a conclusion as to the direct connection existing between the germs found in milk and those that must be contained in the food. And, in fact, such a connection can be traced all along in the milk and more so in the products therefrom, particularly when a change of feed occurs or when fodder is fed which is filled with acid or fermenting organisms, such as wet brewers' and distillers' grains, spoilt ensilage, musty hay, mouldy grain, etc. Practical dairymen know perfectly well what evil effect spoilt or badly kept fodder of every kind has on the quality of the milk and its products. The bedding also on which cows lie or stand has an influence on the bacteriological contents of the milk; it will in a great measure depend on the soundness and freshness of the

bedding which is perhaps spoilt by having been
housed in bad condition and containing spores of
mould, rust, smut or other fungus growths. The
cleanest and most unobjectionable bedding in every
respect is *moss peat* (not peat moss). A great influence is also exercised by the more or less frequent
changing of the bedding, because any carelessness in
this respect forces the animals to lie down in the
putrid and fermenting matter.

Very often milk is still further polluted by the unclean hands of the person milking, by insufficient
cleansing of utensils which during the entire handling of the milk are brought into contact with it, and,
lastly, by the dust suspended in the stable air, being
partly dust from the feed and partly from the bedding or the floor. We all know that to a certain
degree this contamination of milk by the above named
matters and, therefore, also by bacteria, cannot be
entirely avoided and some of these are even absolutely
necessary for the extraction of the products of milk,
but the above considerations clearly demonstrate as
does also longtime experience in dairying, that it is
by no means indifferent what degree of pollution is
attained and to which class more especially the bacterial infection belongs.

When we recapitulate all that has been hitherto
said, and consider that all these bacteria possess a
marked altering and changing influence on the ingredients of the milk—some slower, others more rapidly,
and that they assist and stimulate one another in

their mission to decompose, we can easily comprehend how milk that is heavily disseminated with bacteria must lose its keeping qualities and that a possibility of infection by bacteria, which is bound to produce annoying complications in the milk and its products, is by far greater in a stable with chronic filthiness than where methodical care is taken to suppress every cause for such infection.

All of us have repeatedly heard complaints on the lack of cleanliness in the stables as practiced by many farmers; we meet with these complaints in every agricultural journal, in the reports of dairy commissioners, commissioners of agriculture and presidents of creamery associations, but only in Germany have I noticed an effort to bring this degree of uncleanliness more forcibly unto our conception by the uncontestable figures of actual weight. *Renk* found, for instance, an average of 0.015 grammes of cowdung in every quart of milk sold in the city of Halle, of 0.009 grammes in Munich and of 0.010 grammes in Berlin. This gives a total of fifty tons of cowdung per annum consumed by the unsuspecting public of Berlin. There cannot be the slightest doubt but what the same state of affairs prevails in this country. The number of bacteria found in milk gives a fair scale to measure the cleanliness by, but this is the case only when investigation closely follows the milking. *Cnopf* found from sixty to one hundred thousand germs in one tenth of a cubic inch, and *von Freudenreich* found from ten to twenty-five thousand.

A perfect condition of the milk is not merely dependent on the cleanliness while drawing it, but also on the carefulness with which milk is kept after milking. It is easily understood that unclean vessels and utensils are able to infect clean milk with bacteria, and that an infection with these will unavoidably follow if milk is left standing, for any considerable time, in the air of the stable impregnated with bacteria. The greatest influence on the number of bacteria is, however, exercised by the temperature to which milk is exposed after milking, as the vitality of bacteria is greatest at bloodheat and somewhat above that.

The number of germs will, according to *Weigmann*, multiply:

a. at 95° F. (Bloodheat.)		b. at 60° F. (Cellar temperature.)	
After 2 hours	23 fold	. . .	4 fold
" 3 "	60 "	. . .	6 "
" 4 "	215 "	.	8 "
" 5 "	1830 "	. . .	26 "
" 6 "	3800 "	. . .	435 "

We see from the above that not even the temperature of the cellar is able to prevent these germs from propagating, although for the first few hours they are considerably restrained from so doing. The preservation on ice has a far better result—a number of observations made were unable to detect any increase worth recording.

It is sufficiently clear from these numbers that

temperature exercises an enormous influence on the
propagating powers of bacteria and explains the fact,
so widely known, that milk which is at once cooled
after drawing keeps much longer than uncooled milk.
This influence is so great that even a very cleanly
drawn but insufficiently cooled milk is apt to contain
more bacteria and spoil sooner than a filthy milk
very strongly cooled.

CHAPTER III.
Decomposition of Milk.

We saw a short while ago that all decomposition of organic matter is to be attributed to the influence and activity of bacteria, and when we see that milk, soon after having been drawn, may contain such enormous numbers of bacteria, it is not to be considered strange that it should soon spoil. The first noticeable act of vitality of these inhabitants of milk is generally the souring of the milk, *i. e.*, the transformation of milk sugar into milk acid. A considerable number of such bacteria are now known which cause this transformation, and we know of them further that they have only this effect and no other. In the course of this milk acid fermentation, as we often hear it called, not all of the milk sugar is transformed into milk acid but only a certain part of it ; in other words, a certain amount of milk acid is only formed and after its formation the fermentation or transformation comes to a standstill. Bacterial life has ceased to make itself felt, or, to use the expression of the renowned French scientist, Pasteur, "the acid ferment (*ferment lactique*) has become latent."

The forming of milk acid is, then, the cause of the casein, the most important of the albuminoids of milk, being liberated from its affinity with lime, and the milk "curdles." This kind of curdling is essentially

different from other forms of curdling of milk, which
are partially based—similar to the acid curdling—on
the action of a living ferment, the bacteria ; partially,
however, their appearance is due to the action of a
dead or so called chemical ferment.

The best known curdling is the one accomplished
by rennet which is a chemical ferment. By this process the casein of the milk is chemically changed,
inasmuch as it is transformed after separating the
"whey protein," a peptonic matter, into so-called cheese
or, as we often call this albuminous matter, into paracasein.

This rennet curdling is similiar to another curdling
of milk, which must be laid to the action of certain
bacteria and which envolves a simultaneous transformation of the casein. Certain bacteria are able to
cause a ferment to exude, which acts similarly to
rennet on milk, forcing it, without previous acidulating to a rennet-like coagulation ; however, in most
cases this "bacterial rennet," as we might call it,
seems to have the effect of again dissolving the
formed cheesy mass and transforming it into a soluble
matter—"peptonising the albumen," as the scientist
would call it. This bacterial ferment, therefore, be-
haves quite differently from the rennet ferment which
does not have the dissolving power. It is, however,
not excluded that these bacteria may, at the same
time or later, separate a second ferment which posesses
this very effect to a certain degree.

Now, raw milk at all times contains such bacteria

which tend towards its being curdly, be it either acid
or rennet curdling; in most cases the acid bacteria
predominate in numbers, or, at least, their activity is
more readily noted. Aside from this acid-curdling,
and dependant on the proportion of the acid bacteria
to the rennet bacteria, we find that a rennet curdling
is going on later, simultaneously or even sooner, and
which, in most cases, is not noticeable because the acid
curdling has already been completed. Only in the
case where the number of rennet bacteria predominate by far, we see a curdling without previous acidulating which happens in the "cheesy milk." These
rennet bacteria—which are also commonly called
butter acid bacteria, because they generally possess
the property of producing butter acid—play an important part in the keeping qualities of milk. While
we find it easy to counteract or retard the milk acid
fermentation, and thereby the acid curdling, we shall
see that it is connected with considerable difficulty to
avoid the rennet curdling by bacteria.

From the foregoing, the reader should receive the
impression of the great importance of producing a
milk containing the smallest possible number of bacteria, as upon this depends the success of manufacturing it into normal infants' milk, and, for this same
reason, it has been found unrecommendable to separate the agricultural part, the production of the cow's
milk, from the technical part; the treatment we shall
describe later on.

No manufacturer of infants' milk, no matter what

name it is sold under, can conscientiously guarantee
the pureness and healthfulness of his milk unless he
has had personal supervision and control of the physical condition of the cows, the food they have eaten
and the treatment they have received.

Methods of Preserving Milk.

As we have seen in the foregoing, the changes in
milk, more especially its curdling, are due to the
action of bacteria (and to some other fungus spores),
we shall, therefore, succeed in preserving it if we can
either defer the action of the bacteria or remove them
entirely. Both methods have been tried for some
time. Efforts have been made to prevent the impending souring by adding chemicals, the curdling
by so-called preservalines, and also to counteract, by
refrigerating, these phases of commencing decomposition; but of late all efforts have been directed towards killing the bacteria themselves through the
application of heat, so as to secure in this manner the
keeping qualities of milk even for a longer period.

CHAPTER IV.
Preserving Milk by Chemicals.

I have hesitated for some time to say anything on this subject, because the preservation of milk by chemicals, even if it were justifiable to practice it, is not a procedure that in any manner or form should be contemplated by those for whom I write, nor is it in any way conducive of better results towards attaining a milk with keeping qualities sufficiently pronounced to serve all requirements, as the methods which will anon be treated, such as cooling, Pasteurizing and sterilizing, and which are now conceded, and justly so, to be the only methods which should lawfully be countenanced *anywhere*. Yet when I reflect that it is only by exposing the misuse of chemicals for preserving milk that a chance will offer itself to dwell on the pernicious results which may follow, it will be accorded that it may be best to show all there is in it.

Of the many and most frequently used ingredients which have been adopted by the smaller retail milk dealers, and are still used, to prevent or cover the impending souring of milk (and often in the erroneous supposition of retarding it), none are more generally used than soda. By its admixture it is brought about that the milk acid, formed from milk sugar by the action of acidulating bacteria, is dulled and, consequently, not perceptible to the organs of taste. During this process the multiplication of germs in

4

the milk has not been counteracted or suspended, but has, on the contrary, been favored.

Bacteriology has taught us that an alkaline reaction is extremely conducive to the welfare of bacteria, therefore the addition of this chemical may for several hours disguise the acidity, but in no manner will it retard the curdling, with which end in view it has probably been added. Milk treated with soda and kept at a temperature of 80° F. will keep from becoming sour for from twelve to twenty-four hours; at 95° F. for from six to ten hours, while the curdling, however, has by no means been retarded.

A simple experiment will show that the curdling sets in at about the same time in samples of pure milk and in such treated with soda, if kept at the same temperature. As the beginning of curdling in all pure milk is nearly entirely dependant on the quantity of milk acid formed therein, it would seem at first sight as if this result were contradictory. We have, however, seen that the curdling of milk is not only enacted by such bacteria, which produce acidity, but also as well by a large number of other species of bacteria which have the faculty to produce a rennet-like ferment. By a low alkaline reaction the propagation and multiplication of bacteria in milk is favored and, therefore, also their effect, so that the dulling of the acid is compensated by the more rapid development and increased activity of the rennet producing bacteria. For this reason the result of such investigations depends largely on the quantity of rennet producing bacteria contained in the milk. If we now try to find out which bacteria are of the rennet producing kind, we shall see that they are principally those that live in the uppermost layers of the soil and have been collected with the hay and other fodders, so that we may presume that such milk which has taken up many bacteria in the stable, or which has been strongly polluted after having been drawn, will more rapidly advance toward rennet curdling than milk which has been less infected.

Among other ingredients used, presumptively, for the preservation of milk are lime, borax, boracic acid and salicylic acid. Some of these are even now used extensively and have been for many years, for instance, by the farmers of the North Sea coast, because for them it was a matter of existence to keep their milk sweet for at least thirty hours to enable it to reach their only remunerative market which, to the greater number, was London.

Investigations on the preserving merits of boracic acid, common salt and salicylic acid show the following results :

Admixture.	Commencement of Acidity Confirmed by Tasting.	Commencement of Curdling.
0.02 per cent. boracic acid.....after 30 hours....		after 47 hours
0.04 " " ..... " 35 "		" 47 "
0.06 " " ..... " 56 "		" 60 "
0.02 " salt.... .... " 26 "		" 30 "
0.04 " " .... " 26 "		" 32 "
0.06 " " .... " 26 "		" 32 "
0.02 " salicylic acid.... " 33 "		" 58 "
0.04 " " ..... " 47 "		" 82 "
0.06 " " ..... " 144 " ..		{ was not curdled after 8 days
Pure milk.................. " 25 "		after 28 hours

Judging from the above, table salt can hardly be called preserving, while boracic acid is considerably so, and salicylic acid even more so. With the latter it is quite noticeable that it prevents the curdling for an extremely long period.

In regard to the difference of taste produced by these preservatives, the admixture of boracic acid and of common salt are hardly to be detected, but that of salicylic acid very plainly, as it gives milk a sweetish taste. The preserving effects of these admixtures was found lessened in proportion to the time which elapsed between milking and that of adding the chemicals, a natural conclusion when we remember how rapidly the germs multiply.

A sample of a "trebly concentrated preserving salt," manufactured at Stuttgart, Germany, was ascertained to be composed of salt and boracic acid, and an admixture of it in the strength of 0.068 per cent. added to milk had a preserving effect of 24 hours. *Soxhlet* also investigated the preserving qualities of boracic acid and found that curdling was protracted for:

35 hours by an admixture of 0.1 per cent.
65   "   "   "   0.15   "
147   "   "   "   0.2   "
231   "   "   "   0.4   "

Temperature, as well, has a most important influence, and milk with an admixture of boracic acid (1 gramme to 1 liter) was kept from curdling for 30 hours, if kept at a temperature of 60° F. or below, and that even half of this quantity of the chemical was able, at the same temperature, to preserve milk for 21 hours longer. But the value of a preserving chemical must not only consist in protracting the curdling of raw milk, but also in preserving it in such a manner that it will not curdle when being boiled. The curdling at the time of boiling could be protracted for :

10 hours, by an addition of 0.05 per ct. boracic acid.
33   "   "   "   0.01   "   "

Yet we should never lose out of sight the prime requisite to be demanded from all milk and, therefore, also from preserved milk : it should be absolutely healthy, and this cannot be upheld, even in the face of statements made by eminent scientists who teach the contrary and who claim that these perservatives are harmless or have no deleterious influence whatsoever. When we reflect for a moment that the public buys our milk " bona fide," intending to use a great part of it for the nourishment of infants whose tender stomach we may compare to a highly tuned and sensitive instrument, whose cords connect it, as it were, with the entire nervous system, the brain, the heart, in fact with the aggregate vitality, that for these infants even the purest cows' milk is an absolutely unfit diet, we should find no hesitation in arriving at the conclusion that every tampering with the milk in the hands of the farmer or the dairyman, by the use of chemical admixtures, is little short of criminal. Formerly great efforts were made to establish the harmlessness of boracic acid, but more recently it has been repeatedly proven that it has a deleterious influence on the mucous membrane of the intestines, even if administered in doses such as we have seen are necessary to be added to milk; this acid has been used not only in milk, but in a large variety of foodstuffs and fluids. Consumers would after some time be troubled with salivation, increased urination, diarrhea, loss of weight and on several occasions in aged persons—death insued.

From Norway and Sweeden, where the use of boracic acid seems to be quite prevalent, more so at least than anywhere else, repeated cases of poisoning by the comsumption of such " preserved " milk have been reported. In other countries the use of this acid as a preserving chemical has been entirely condemned. Also in regard to salicylic acid it has been established that, even in the minutest doses, its continued use is harmful to the entire human organism, more especially to the nervous system, and the French sanitary authorities are wageing a lively war against its use as a preserving chemical in the manufacture of canned and bottled foodstuffs. Equally obnoxious is the admixture of bicarbonate of soda to sour milk, because it has a laxative effect and should certainly not be tolerated; the same may be said of benzoate potash, hydrogen peroxide and ozone; even if inoffensive in a pure state the trouble here remains

in the fact that they seldom can be procured in that
state.

The final conclusion regarding the use of all these
chemicals is that milk may be preserved for several
hours by using them, but we also see that the preserving action of these salts is not considerable, so
that not much is gained. For this reason their use
has not become extensive, particularly in cases where
milk was to be preserved for several days. As a
whole, their use has up to date, been limited to the
small milk trade, and all efforts to generalize their
adoption which are at present made, or may be made
in the future, should find a timely end by the promul-
gation, among farmers and dairymen, of more efficient
and harmless ways of preserving their milk ; by the
instruction of the consuming public as to the dangers
of polluted milk, and by the enaction and enforcement
of laws and ordinances, in all States and communities,
which shall tend to protect the entire population
placed under their care from injuries through milk
polluted by chemical admixtures, and therewith prevent the lives of millions of infants being left at the
mercy of unscrupulous greed.

By far more recommendable than the chemical sub-
stances are those expedients which strive to impede
action and multiplication of bacteria through influences of temperature, and which have been known
ever since the most ancient times viz.: the cooling
and the heating of milk.

CHAPTER V.

Preservation by Cooling.

From the experiments previously noted, it will have become clear what influence temperature has on the propagation of bacteria, and this influence is so much stronger inasmuch as the temperature can be lowered, and, naturally, it was not long before attempts were made to ascertain the keeping qualities of *frozen* milk. In some cases this expedient is resorted to where milk is to be preserved for long journeys. A part of the milk supply of Paris, France, is brought to town in this form, frozen by machinery in vessels with elastic sides and then thawed out before consumption. It is reported that this milk does not differ either in appearance or in taste from fresh milk, and that it can be worked into the products of milk with good results. Also on board of some of the trans-Atlantic steamships frozen milk has been shipped for use for years. This milk is first treated in a refrigerator, and then frozen. The freezing of milk, however, has one serious disadvantage, which consists in the disintegration of milk during the freezing process, which, notwithstanding the previous refrigerating, consumes several hours of time, and, consequently, the cream separates. This frozen block consists of skim milk, on which there is a layer of cream, while in the middle of the block a funnel shaped cavity is formed, which contains unfrozen, but very concentrated milk.

Vieth, of London, has experimented with such frozen milk, and found the quantity of cream 8.8 per cent.; the skim milk 64.7 per cent., and the fluid or unfrozen part was 26.5 per cent. The chemical analysis gave the following results:

	Ice or Frozen Part. Cream.	Skim Milk.	Unfrozen or Fluid Part.
Specific weight. . . .	1.0100	1.0275	1.0525
Water.	74.44	92.10	80.54
Fat.	19.23	0.68	5.17
Albumen	2.64	2.80	5.38
Milk sugar.	3.33	3.95	7.77
Ashes.	0.52	0.60	1.18

We remark that while the disintegrating action separates the fat and allows it to freeze by itself, the other constituents—ashes, milk sugar and albuminoids—remain in about equal proportion to one another. But it is this very circumstance, the separate freezing of the milk fat, which is disagreeably conspicuous in frozen milk, because the cream does not again mix so completely after having been thawed out, consequently the milk does not present the homogenous fluid that there was before it was frozen.

The analysis of *H. D. Richmond* found the frozen part to contain 96.23 per cent. of water and but 1.23 per cent of fat.

If circumstances do exist under which frozen milk

may be looked upon as a desirable commodity, or which hold out a prospect of widening the circle in which fresh milk may be utilized, they must, however, not be looked for in connection with the manufacture of infants' food, because it is not merely the above mentioned disadvantage of separating the cream, but in frozen milk the bacteria are yet alive, though dormant, and ready to resume their work of

ARCTIC COOLER.

destruction as soon as they are again brought into congenial temperature. We must ever bear in mind that in the manufacture of milk for infants the keeping qualities are of value only when accompanied by absolute freedom from infecting germs of all kinds, and that the process of freezing is merely a mechanical means of stopping the activity of bacteria and in no way able to correct any physical defect the milk may have posessed before the freezing. For these reasons the call for frozen milk has ever remained a limited one, while the process of merely cooling milk is one of the utmost importance, as we shall later see.

CHAPTER VI.

Preservation of Milk by Heating.

We may suppose that the custom of preserving milk by heating is as old as the cow and the use of the fire. The simplest way to accomplish it is the one in practice in all households over the whole world wherever fresh milk is to be had : the boiling of it in an open vessel, and its subsequent cooling. Milk-boiling pots have been introduced to avoid the boiling over and the consequent disagreeable smell and loss of milk, but we can not go into a discussion of their merits and failings. The necessity, or the wish to preserve milk is, however, not only a desideratum for households but by far more urgent for dairies, more particularly for such dairies that return the skim milk to the patrons, but also for dairies that have milk routes in cities and for the whole milk trade in general.

It is well known to all who are in any manner connected with or interested in the milk trade, how difficult and dainty an article milk is, on account of its easy decomposition, in all cases where it has to be brought to town from great distances and from localities that could not command the use of refrigerating appliances during the transit. One of the first steps taken towards attaining greater security was simply

the boiling of the milk in large kettles, imitating the
process of the households. In this way one could
well obtain a longer keeping quality of the milk of
from 12 to 24 hours, but there was the disadvantage
to be contended with that the boiled taste is not liked
and damages the sale, although it is uniformly the
custom to at once boil the milk when bought. This
is quite a peculiar difficulty encountered everywhere,
which is, perhaps, accounted for by the distrust felt
towards boiled milk and the preference given to the
raw article and, perhaps, not without good cause ; on
the other hand it is positively a fact that by a
majority of consumers the taste of boiled milk is not
liked, and it may readily be conceded that the specific
agreeable taste of unboiled milk is everywhere pre-
ferred to the former. Besides, it was found that in
following the way just mentioned of boiling the milk,
the addition to its keeping qualities, was entirely
too short to be of any considerable benefit even for
the closer markets, and that not much could be
gained unless the milk could by boiling be preserved
at least for a couple of days, or, if possible, to give it
an undefinite durability. Trials in this direction
seem to have been instituted soon after science had
instructed us as to the real causes of decomposition of
foodstuffs, and pointed out the path in which a remedy
might be looked for. The pioneers in this line of
work seem to have been Pasteur and Appert, although
their investigations did not lead to a single success, if
we may judge from the very transient notoriety which

their " preserved milk," as it was called for some time,
acquired.

The next great success in this work was to fall to
America, by Gail Borden's invention of condensed
milk, whose innumerable disappointments, however,
may well be taken as a measure of the difficulties to
be encountered by every advancement connected with
the preservation of this, the most necessary of staple
foods of humanity. And it is, perhaps, as well that
it should be so. Condensed milk, as it is manufactured to-day, with and without the addition of sugar,
is come to stay among us because it has the great advantage of being reduced in bulk, of reducing the
cost of packing, and is a great saving in freight for a
comparatively large quantity of milk ; besides, it can
be kept in excellent condition for a very long time.
The change in taste has, naturally, not been avoidable
because even the milk condensed, without the addition of sugar, has the smell and taste of over-heated
milk, and a slight reddish hue.

After establishing this " condensed milk " a number of other more or less "condensed" milks appeared
in the market, but with little success as infants' milk ;
they have disappeared (with the exception of one
or two brands) as they could not compete with the
superior uniformity of excellence in the Borden milk
and had against them the brownish color of their product.

Condensed milk is to-day recognized as a boon and
a blessing the world over, its production and manu-

facture although highly interesting is, however, an industry by itself, a description of which we cannot here enter into.

There had, in the course of time, been a distinct parting on the roads pursued by experiments and investigations both purporting to lead to the best method of preserving milk by heating. Some advocated a short heating at temperatures under 212° F., others operated at temperatures over 212°. In course of time the first method was called " Pasteurization," in honor to the French scientist Pasteur, because this celebrated investigator had first adopted the heating of fluids, particularly of wine and beer, to 140° F. as a means for their preservation. The other method, that of applying higher temperatures, was named Sterilization, because the milk was, apparently, made *sterile*, that is to say: the milk was freed from the micro-organisms it contained, by which process alone it is possible to attain an unlimited keeping quality for the milk.

CHAPTER VII.

Pasteurization.

In some dairies, as we have seen before, the habit of pasteurizing in common open kettles had been in use. The next step was, the heating of the milk in tightly closed kettles, when an enormous improvement was at once recorded. The clumsiness of the first apparatus and the desire to combine the milk-heater with the action of the cream separator were the cause of a large number of inventions of different apparatus which may now be found in a large number of dairies. The first of these apparatus dates back to 1882, when it was patented by Albert Fesca, who termed it "a continuously working apparatus for the preservation of milk by heat." It would be useless to attempt to describe all these different inventions, many of which were used for a very short time, and it will suffice to give the principle on which it was claimed they performed the preservation of milk.

An upright cylinder of galvanized copper, and surrounded by a closely fitting steam-jacket, contained a stirring arrangement by which the milk, that entered from below and was forced out through the top, was kept continuously moving so as to avoid its scorching at the sides close to the steam-jacket. All these apparatus, however, had, and have yet, some defects in common : one is the aforesaid burning or scorching of the milk, and another the great insecurity of attaining the desired degree of heating for all the milk passed through the apparatus. As the injection of the milk was continuous it was unavoidable that some part of the milk would at times rise and find the exit without having attained the prescribed degree of heat. As we may suppose all such milk heated to 165° or 170° acquired the taste of boiled milk, a defect which, it is safe to say, has hardly a chance to be overcome. The great heat that has to be kept up on the metal sides of the copper cylinder containing the milk is one of the great defects of all of our present pasteurizing machines, and it is certain that this must be remedied before pasteurization will become an operation of universal practice. After what has now been said there would be justice in contending that the present pasteurizing apparatus will be even less successful if temperatures of not more than 176° F. can be applied. This will hold good only for the present apparatus; in other words, all these apparatus have a defect, and a signal defect at that, which involves the scorching before mentioned. This great defect is that the milk is heated for too short a time and that it remains inside of the apparatus for too limited a duration, consequently necessitating a comparatively excessive heating at the sides of the milk to attain an enhanced keeping quality.

From this reflection and from the observation that the " boiled " taste of milk is already noticeable at

temperatures of 165° to 170° F., it must be concluded that the application of a temperature under
170°, but during a more protracted period, must be
the right thing, and experiments accordingly made
have confirmed this conclusion.

We know that all changes which take place in
milk must be traced to the presence and activity of
spores, ferments, etc. We must conclude herefrom
that the keeping quality of milk is dependent upon the
quantity of such germs contained therein, and that
also the success of pasteurization must depend on the
efficiency with which it has killed the majority of
germs or not. If we, therefore, wish to study the effect of heating on the durability of milk, we have to
study the effect which heating produces on the milk
fungi, and such experiments have to be carried on by
purely bacteriological methods, which in their sim-
pler forms we shall have to adopt when testing milk
to be prepared for infants' food; a closer description
of the apparatus used will be brought in the chapter
treating of the manufacture of artificial mothers'
milk.

The defects attached to pasteurizing apparatus
have been clearly demonstrated by a large number
of experiments. It has been proven that certain bacteria which had been introduced into the milk, for
instance, bacteria of tuberculosis, can be killed at a
temperature of 154° to 155° if they are only exposed
to this temperature for about thirty-five minutes.
From this it was correctly concluded that other bac-

teria, more especially those commonly contained in milk, could be killed at a temperature as low as 176° or even 167°, if only they could be kept in this temperature for a sufficiently protracted period. This conclusion having been reached and confirmed, it was at once plain that the apparatus to be used would have to abandon the aim of continuous operation and adopt the principle of periodic filling and emptying. In his exhaustive researches in this direction, *Bitter* reached most conclusive results. Beginning again with milk to which bacteria of tuberculosis were added, he heated this in an apparatus of his own invention to 154° F. for fifteen, twenty and thirty minutes respectively, in separate lots. Corroborating not only the result of his previous experiments in the laboratory, which had shown that thirty minutes were sufficient to kill these bacteria exposed to 154°, it was found that even half of this period, fifteen minutes, sufficed to attain the same result. After this the experiments were extended to examine the effects of pasteurizing on the ordinary bacteria of milk under varying degrees of heat and varying periods of exposure to such heat.

It was of the greatest importance to attain a standard of comparison, not only for the preservation of the milk, but also as to its fitness for consumption. The investigations were, therefore, extended to the appearance, smell and taste of the milk treated, and to detect every change in these properties on which the value of milk as an article of consumption so largely

depends. It was equally of importance to establish
a method to enable an examination of the keeping
qualities of milk which would manifest the spoilt
character of the milk even before this should be ex-
ternally visible.

Commonly, the keeping quality of milk is judged
by the earlier or more protracted appearance of curdling. But milk is really spoilt before this occurs, as
the requisites for curdling are all present, so that it
needs only a slight warming to effect the separation.
The curdling of milk is, however, generally the consequence of its acidity, and one would believe that
the reaction of the milk should furnish a measure for
the expected appearance of curdling. In the case of
raw milk this measure could, perhaps, be adopted,
and, in fact, experiments have recently been made to
determine what must be the degree of acidity to make
milk curdle at warming; this will be described later
on. The method, even if reliable results are to be
obtained by it, is one of complicated manipulations
suited only to laboratory work, and has for this reason
not received the attention and application it merits as
a means to examine milk brought to market, which
in itself is a most desirable investigation. When it,
however, comes to the manufacture of milk into food
for infants we can not operate with any such uncer-
tain factors, therefore the degree of acidity in the
milk to be used for this purpose must needs be ascertained by the manufacturer; there must be, absolutely,

no item in the entire process left to haphazard or to chance.

We have previously seen that, besides the acidity, there are other causes for the curdling of milk, that the latter may even curdle without being at all sour, and that there exists a large number of bacteria which possess the property of separating a rennet-like ferment and which, consequently, if they be present in sufficient numbers, are able to make milk curdle. Milk in which such bacteria predominate will curdle very easily at warming without any abnormal degree of acidity having previously been observed. The reaction of milk is, therefore, not always an unerring sign of probable curdling when warmed, but the warming, itself, rather constitutes the surest experiment towards the examination of milk in this direction, more particularly of such milk which is produced under conditions entirely remote from our observation. This is also true of pasteurized milk. All bacteriological investigations of pasteurized and sterilized milk have shown that it is more especially the group of rennet—or butter acid bacteria—which in their endurate form of spores resist the influence of heating better than other bacteria. For this reason well pasteurized milk contains, when it becomes older, principally these bacteria, and it may curdle in the course of time *without* perceptably increasing in acidity.

The keeping quality of pasteurized milk can, therefore, not be examined by the chemical reaction, but rather by the direct experiment of curdling : it must stand warming without curdling, because on this the whole value of the milk, not only for the household but also for the manufacture into its products, is dependant. It has been established that milk heated to 154° and kept there for thirty-five minutes retains but very few bacteria, that the pasteurization was as complete as can be attained by any heating under 212° F. The length of time which such pasteurized milk keeps was found to be from six to eight hours longer than non-pasteurized milk of the same date and both kept at a temperature of 86°, at least ten hours longer at 77° and from fifty to sixty hours longer if kept at 65° F. This enhanced keeping quality may also be regarded as constant and not varying. The time of heating, namely, thirty-five minutes, had been retained because this had been found sufficient to kill the bacteria of tuberculosis, frequent extraction of samples during the process had shown that already after fifteen or twenty minutes none had remained alive, so that a duration of heating for thirty minutes, consecutively, at 155° can be pronounced, under all circumstances, as a thorough pasteurization. Further experiments, with a higher temperature, were made with skim milk, when it was found that 167° kept up for fifteen minutes was entirely sufficient.

Here the taste of the milk was hardly altered, although the temperature was nearly up to where albumen coagulates, and therefore a change in taste could be expected. It was, therefore, surmised that full milk would stand heating to 167° equally well without acquiring the boiled taste, and experiments have confirmed this supposition. The keeping quality of a milk pasteurized at 167° was enhanced by twenty-four to twenty-eight hours if the storing temperature was 73°, and sixty hours if the temperature of storage was 60°, and was also enhanced in the same measure as by a pasteurization at 155° lasting thirty minutes.

The investigations of Prof. H. L. Russell, of more recent date, have thrown a great deal of light on the effect of pasteurizing on the different species of bacteria in milk. Excluding from consideration those species that have occurred only sporadically in the cultures of bacteria, fifteen different forms in all have been isolated from normal milk and cream. Of this number, six different forms have predominated in a large degree. When classified as to their effect on milk they are grouped as follows :

Species producing lactic acid............................. 3
Species causing no apparent change in milk................. 7
Species coagulating milk by the production of rennet and
    subsequently digesting the curdled casein.............. 5

In the same milk, after pasteurizing, only six species were isolated. Of these, three had no apparent action on milk, while the remaining three species curdled the milk by the formation of rennet and then subsequently digested the same by the action of a tryptic enzyme. The lactic acid producing species that make up the majority of individual germs in the raw material were entirely destroyed by the pasteurizing process. This class, as a rule, does not form endospores, consequently they are unable to resist the heat employed in pasteurizing.

In the normal milk it is to be noted that while the majority of individual germs belong to the lactic acid producing class, yet a larger number of species producing little or no acid are to be found in milk. These are, doubtless, the organisms derived from extraneous sources. They are germs associated with dirt and excreta, and gain access to the milk during the milking. Baccillus mesentericus vulgatus, the common potato baccillus, was frequently isolated from the pasteurized as well as from the raw milk. As these organisms that are thus associated with filth of various kinds are able to persist in pasteurized milk by virtue of their spores, it emphasizes the well-known lesson that scrupulous cleanliness is an absolute essential in dairies that pasteurize their milk for direct consumption. Cleanliness in milking diminishes materially the amount of this class of bacteria that gains access to the milk. The lactic acid bacteria, those that are essentially milk bacteria by prediliction, are the forms that are habitually present in the milk duct. These are the bacteria that cannot well be kept out even by the greatest care. They are, however, the forms that succumb most easily to the pasteurizing process.

In reviewing these results it may seem singular that the duration of keeping qualities of pasteurized milk, particularly at higher temperatures, is not very much greater than that of non-pasteurized milk, so that the result does not seem to be very encouraging. But we must remember that milk is seldom exposed to such a temperature as 73° in the longest transits. Therefore, if properly cooled before transportation and the most common precautionary measures are observed (such as keeping some ice near the cans or using refrigerator cars) results will generally prove satisfactory. It will be readily comprehended that milk will keep so much better after pasteurization the more rapidly and strongly it is cooled after heating. The larger the transporting vessels are the more easily will the temperature be kept down.

If we now consider all conditions, it may be stated with certainty that the keeping quality of properly pasteurized milk will be thirty hours, even during the hottest summer days, and, at lower temperatures, naturally ever so much longer. A matter of the highest importance, aside from the enhanced keeping quality, is that in such milk cream will rise and become butter just as easily and the butter not have the slightest trace of taste to distinguish it from other butter made of non-pasteurized milk. Pasteurizer and cooler should, naturally, be mounted in a manner to avoid as much as possible the exposure of the pasteurized milk to the air. Pasteurizing machines find the greatest field of utility in creameries where skim

milk is returned to the patrons, and as they are
capable, when properly managed, to disinfect the
skim milk at a trifling cost from the pathogenic—or
disease-producing bacteria—that is, from those that
are apt to carry and spread infectious diseases such as,
for instance, those of tuberculosis, typhus, foot and
mouth disease, scarlet fever, etc., they should be in
general use.   In several European countries—Germany, for instance—the creameries are obliged by law
to make use of them.   When we refer, however, to
the object of this treatise : the manufacture of milk
into a healthy food for infants, it must be said that
the pasteurizing machine does not find an employ-
ment in this process because a higher standard of
efficiency must be aimed at, yet it seemed advisable
to explain the effects of pasteurization so as to be
able, later on, to define the difference between it and
sterilizing, and avoid the confusion that in the
minds of many now exists with reference to these
processes.

CHAPTER VIII.

Sterilizing.

Pasteurizing does not kill all bacteria as we have seen, because either the temperature has not been high enough, or, as is the case in the common apparatus with continuous working, has not acted long enough on the milk, partly because the endurate forms the spores of certain bacteria can well endure temperatures of 212° F., particularly if these are not kept up for a longer time.

Investigations have shown that there exist, comparatively, not a few bacteria that are able to withstand high temperatures; Cohn's investigations have proved that the hay bacillus (bacillus subtilis) will at a temperature of 120° F., at which, ordinarily, other organic life commences to die, still increase rapidly, and Miquel found a bacterium in water, which not only endures perfectly a temperature of 158°, but prospers in it; for which reason it was named " bacillus termophilus." Now, if bacteria are able to resist, even in their vegetative period, the part of their lives in which they, apart from a great display of activity and multiplication, are keenly susceptible to outward influences, to such high temperatures which are commonly considered as the limit of organic life, or, if they ever require such temperatures to deploy their full vital energies, how much greater must then be the possibility that these bacteria will in another, their endurate form, be able to resist such higher temperatures? We know, in fact, quite a number of bacteria whose endurate forms, the spores, are able to endure such intensive heat as would at once kill all other organic life. The baccillus subtilis has been cooked for two hours and a half, consecutively, at 212° and not lost its power to germinate, and another investigator found that this ironclad baccillus could be killed only at 240° of heat. *Globig* found a baccillus living on the potato, the "red potato baccillus," the spores of which could be pronounced dead only after having remained in steam of 212° for six hours, and in steam under pressure at 235° the same spores were yet alive after forty-five minutes.

It will, therefore, easily be understood that in a process like the pasteurizing, which seldom exceeds 160° to 175°, there very frequently remain live bacteria and spores in milk, which are sure to spoil it after a longer or shorter time. The desire, however, to give milk keeping qualities, not only for days but for weeks and months, is an urgent one, and, therefore, all efforts have been concentrated to destroy all bacteria by the application of heat above 212°, and thereby to reach the desired keeping quality. Reviewing the observations hitherto enumerated of the temperatures at which the spores of several of the more resistant kinds of bacteria may be killed, we see that milk which contains, for instance, the wide-

spread and common baccillus subtilis would have to
be heated for a considerable time to 240° to insure
any degree of security of its having been killed.

Pasteur records amongst his experiments of sterilizing milk that the hay baccillus was found killed only
after a heating of several hours' duration to 230°, or
after heating for half an hour to 266° F. To such
excessive heat we cannot, however, expose milk without its palatability being seriously impaired, so that
sterilizing at such temperatures is practically not to be
thought of. We note that in the beginning all these
experiments tended merely to produce a keeping
quality in the milk, and only in the course of time
the expediency became apparent of combining with
it a sanitary amelioration by its thorough disinfection.
We shall first review the effects of sterilizing from
the standpoint of longer keeping qualities, and turn
thereafter to the merits attained by the disinfection.

Among those that entered the occupation of building
sterilizing apparatus, two distinct methods were very
soon adopted—the one heating to high temperatures
and then hermetically sealing the vessels containing
the milk, the other advocating a repeated heating and
intermediate cooling at different degrees of temperature, which is termed "fractionized sterilization."
Tyndall was the first to advocate this method, and
Dahl adopted it, cooling milk first to 55° and then
heating it to 158° for four consecutive times and
cooling the milk to 104° between each heating, the
separate operation consuming one hour and a half

each, and after the last cooling another heating for
half an hour to 212° was given, and then finally
cooled to 60°. This method was, as we readily comprehend, far too tedious to be extensively adopted or
applied, later on it was modified to but two heatings
at 158° and the last heating to 212°, so that only
three heatings in all were given. But even this reduction was not sufficient to bring it into general use,
also the costs of the repeated manipulations were by
far too heavy. It was then reduced to but one heat-
at 194°, a subsequent cooling, and then a final heating to 215°. The manner of putting 'this method
into practical operation was that the milk was filled
into glass bottles with the porcelain stopper and wire
closing arrangement. These bottles had been previously sterilized in flowing steam of 212° for half an
hour. The rubber rings or washers used with these
stoppers were boiled in water and soda until every
particle of taste or smell had vanished; the rings
were now drawn over the porcelain stopper by
scrupulously clean hands, the bottles filled by a bottling apparatus and placed in the sterilizing chest.
This chest was fitted with a patent arrangement for
closing down the wire fastening without opening the
steam chest (the object being to allow the air in
the bottle to escape during the boiling of the milk)
but to seal the bottles hermetically immediately after.
The temperature produced in the sterilizer by the
steam is descernable on a thermometer, which is fixed
in the covering or hood of the chest with the quick-

silver bulb inside in contact with the steam. In
some of these apparatus an electric bell has been
connected with the thermometer in a manner to close
the contact and ring when the quicksilver has risen
to the prescribed degrees of heat ; but as the heating
has to be done very gradually, or a large number of
bottles will crack and burst, the operator's hand is required constantly on the steam valve and his eye on
the thermometer, so that this electrical arrangement
becomes entirely superfluous.

The inconvenience of losing bottles and their contents by bursting was practically overcome by the
immersion of the bottles in a water bath, and the
success of this simple expedient seemed to prove a
lasting one until a singular defect to it appeared,
which very speedily caused the abandonment of the
water sterilization as far as it was applied in the production of normal infants' milk. It was found that
the bottles used in the water sterilization began, in
the course of time, to loose their brilliancy, their surface becoming dull and gritty by the action of minute
particles of lime which were deposited by the boiling
water, and which defied all efforts to remove them by
mechanical or by chemical means of cleansing. Although this dullness of the glass did no harm to the
contents of the bottles, yet it was found impossible
now to control the proper cleansing of the bottles,
simply because they retain a look of uncleanliness, no
matter what sum of exertion has been expended on
their cleansing.

In sterilizing by steam it is necessary that all air
be driven out of the apparatus, because a mixture of
air and steam gives very unsatisfactory results; the
apparatus should, therefore, be fitted with an escape
pipe, through which all air may be driven out and a
sufficient amount of steam may also continuously escape during the entire duration of sterilization, so as
to maintain a circulating movement of the steam inside of the apparatus; this is essential to equalize the
temperature in all parts of the apparatus, for, without such movement of the steam, either the bottles
nearest to the entrance of the steam will be over-
heated or those more remote not attain the desired
degrees of heat. We have seen that a thermometer
is attached to the hood of the apparatus to indicate
the heat of the steam as it fills the inside, enabling
the operator to regulate the flow in such a manner as
to secure a steady rising of the temperature not exceeding 5° F. in every minute. But the temperature of the steam in the apparatus is no indication of
the temperature of the milk in the bottles to be sterilized, and to know which is of the greatest importance. For this reason it is necessary to fix a second
thermometer in the hood of the apparatus, exposing
the scale of degrees outside, whilst the quicksilver
bulb reaches down and dips into the milk in one of
the bottles inside. This bottle, or rather a bottle
with the neck trimmed off, so as to offer a wider
mouthed opening for the thermometer bulb to dip
into, is so fixed on a bracket that the thermometer de-

scending with the hood or cover will exactly dip into
this milk (see Fig. 18), and consequently the read-
ing on this thermometer will give a fair indication
of the degree of heat attained in all the bottles.
When bottles of different sizes are sterilized simultaneously, then one of the largest sized bottles must be
used to hold the thermometer bulb, for we must take
account of the prescribed time for sterilizing from
the time the largest bottles in the aparatus have
reached the desired degree of heat.

Whatever time may have been fixed upon for the
various periods of sterilization or combinations of
alternate heating and cooling, they should, however,
be closely adhered to, as every variance therefrom, or
negligence in this respect, will at once tell on the
keeping qualities of the milk.

Let us, however, bear in mind that all attention
and neatness during the process of sterilization is
wasted and futile, if the milk has not been produced
and handled with the utmost cleanliness, and here,
again, we may observe that it is not so much the
bacteria floating in the air that have to be feared and
guarded against, than those that cling to matter of
every description: vessels, utensils, hands, etc. The
prime object to be attained, after having applied the
proper sterilizing, is the hermetically sealing of the
milk bottles before the outer air can come into renewed contact with the contents. In what degree
this last and most important requisite is attained, depends naturally on the efficiency of the closing ar-

rangement of the bottles, and it was natural that very
soon a large number of patent devices sprang into ex
istence, some absolutely without any value, others
too expensive to find general adoption, and it may be
safely averred that the ideal sealing for milk bottles
is yet a thing of the future. The porcelain stopper
and wire closing arrangement, has grave defects;
those that have the wire ends fixed in holes at the
side of the neck of the bottle can hardly be properly
cleaned, as colonies of acid bacteria become lodged
in these holes from where they are not to be got out.
Many do not close hermetically, the tension of the
wires being unequal, stronger on one side than on
the other; no acid being admissible in the cleansing
of these bottles on account of its liability to corrode
the wire, they are with difficulty kept clean, the
whole wire fixture darkens in the course of time, becomes rusty, discolors the neck of the bottles and imparts to them a filthy, slovenly appearance; lastly,
the wire and stopper, hanging to the bottle, are much
in the way where these bottles are to be used for
feeding the contents to the infant direct after pulling
on a feeding nipple.

The greatest defect, however, adhering to these
bottles, and the one which principally makes them
unfit to be utilized in the manufacture and dispensing
of food for infants, is that neither the manufacturer
nor the buying public are able, by the outward appearance of the bottle or fastening, to detect if the
sterilizing effect has been complete, or if it even has

been so at the time of closing the bottles, if it is so yet at the time of sale or consumption. A bottle of milk with the wire fastening may look all right when it comes out of the sterilizing apparatus, but if there has existed the slightest inequality of tension in the wires, and the stopper sits one-sided, or with the pressure drawn to one side only, then, when cooling the reduction in the volume of milk, produces a suction strong enough to draw in some of the outer air into the bottle, and with this air, naturally, germs enter. As a consequence, such milk is no longer sterile, but is likely to turn at any time and produce results which, while they may prove disastrous to the consumer, are sure to damage the reputation of the manufacturing dairyman. Several cases of this kind recurring in a neighborhood are amply sufficient to ruin the manufacturer and bring discredit on the article itself. Another porcelain stopper, made by *Timpe*, abandoned the wire locking and trusted to the atmospheric pressure to do the sealing; this would work well and neatly as long as the top of the bottle was ground to a perfectly smooth flange, to which the rubber washer would adjust itself snugly, but this bottle did not find extensive application—firstly, because it was too expensive and, secondly, because during sterilization the expanding gasses from the bottle frequently lift the stopper and washer, which then do not settle down again to their place, so that such bottles have to be readjusted and go through the sterilizing process again.

It should be understood that it is the manufacturer's most urgent interest to offer in the market only such bottles that will plainly show by an outward and infallible sign that their contents are in perfect condition, and this sign must be one easily recognized so that the consuming public will learn to look for the recognized mark when buying milk. *Soxhlet* was fully convinced of this necessity, and constructed an automatic rubber sealing, which works well enough when used only on the small sterilizing apparatus constructed for family use, where the bottles, after sterilizing, can be handled with care, but in production on a larger scale where bottles have to be sent long distances and be exposed to shaking in cases or boxes, the Soxhlet rubber seal is quite unreliable; besides, the mouth of the bottle has to be ground into a concave, which operation raises the price of the bottle to a figure which places it outside of consideration for general adoption. *Stutzer* invented another automatic sealing stopper, which, although it sits firm and works well, is so misshapen as to be most difficult to clean, also its price is about three times as high as what can be allowed for an automatic sealing device.

The requisites demanded from a bottle to undergo sterilization and for holding infants' milk may be summed up in the following points:

The material must be absolutely crystal clear, so that imperfect cleansing may be easily detected; it must be free of air bubbles, and, in manufacturing, must be very gradually cooled to produce a non-brittle glass.

The best color for the bottle is none at all, but light hues of color may be admitted if required for distinguishing the different grades of milk. The shape should be conical and running gradually into the neck, avoiding the bulging out at the neck common to medicine bottles. The inside surface of the bottom should be well rounded towards the sides, so that no sharp furrow may exist inside for any sediment to stick in.

Every bottle with a flaw or bubble should be rejected, as this will make it burst at sterilizing; the glass should not be too thick or heavy, and no lettering of any kind should be moulded into the face or sides of the bottle, because these raised letters obstruct an equal contraction whilst cooling and thereby cause it readily to burst. The neck of the bottle should be of equal width in all sizes used, so that the same feeding nipple may be applied to all. The stopper must be an automatic sealing one, that is, it must allow the air and gasses which are driven out by the boiling to escape without lifting or moving the stopper, so that as soon as the pressure from the inside relaxes the stopper shows sufficient adheasiveness to close firmly around the mouth and neck, excluding the outer air; in fact, it must sit on so firmly as to exclude all possibility of being shaken or pushed off during transportation, but must yet allow of perfectly easy removal by hand. Such a stopper can naturally not have the shape of a plug, but is a hood or cape of the simplest outline, as seen in Fig. 20, yet afford

ing the greatest facility to be turned inside out for
the purpose of cleansing. The only disadvantage of
such a stopper as compared with the porcelain and
wire arrangement is that it is more liable to get lost
or mislaid.

After having taken every precaution to make the
process of sterilizing effective, we naturally evince a
desire of acquiring a knowledge of the degree in
which we have been successful, and this desire becomes an absolute necessity when we turn to manufacturing milk into food for infants.

As by sterilizing, we have given the milk good
keeping qualities, we may keep the milk stored in a
cool place until the investigation which we shall have
to institute is concluded, and shall have shown us just
how long the milk, which we have sterilized at a certain date, will remain pure and sweet if kept at a temperature of 60° F. or below.

The apparatus which we make use of (termed a
thermostat) is an incubator constructed expressly for
the purpose of hatching bacteria or breeding certain
of their species; its outward appearance and constuc-
tion are shown in cut on opposite page representing
a machine built by F. Sartorius, in Gœttingen, Germany, where it is extensively used, and has been
found entirely reliable. There is a heating chamber
in the center with glass pannell-clad door which may
be darkened by prefixing a felt pannelling. Bacteria
grow more rapidly in the dark. This chamber is
completely encased by a water chest, w, the inner sur-

face being of corrugated metal sheathing, so as to
present a larger heating surface. The filling of this

Fig. 8—THERMOSTAT.

waterchest is through a small tube, a, with distilled or
rain water. Enveloping the water chest is a space filled

with isolating material; at k we see the automatic
regulator, an exceedingly sensitive and ingenious arrangement, registering changes in the temperature of
one-fifth of one degree; t, is the thermometer; b, d
and l, is an arrangement for supplying moist air to
the heating chest; o, is the ventilating chimney; c,
m and s, the heating apparatus, coal oil or benzine
being used in the lamp. Now, from each days production of sterilized milk we retain two sample bottles,
and pasting a label on the side of each bottle, record on
it the date of sterilizing and grade of milk contained in
the bottle. The bottles are now placed in the heating
chest of the thermostat and the regulator set to maintain 95°, F., which is the temperature most propitous
to the propagation and multiplication of bacteria.
Morning and evening these bottles must be taken
out, their contents shaken and attentively investigated as to any change in their condition. If any
bacteria or their spores have escaped the effects of
sterilization then they will speedily be brought to
development and their action on the milk noticeable.
The time, therefore, which milk will keep in un-
changed condition in this incubator is a fair indication
of how long such milk will keep in good condition
when kept at lower temperatures. Milk that will
keep perfect in this brooder for twenty-four hours is
likely to keep perfect for one week at 60°, or below,
and milk that keeps for eight days in the chest without curdling will, undoubtedly, keep good for eight
weeks if kept in an ordinary cellar, and ever so much

longer when cooled with ice. This testing should be carried through most strenuously if one would avoid disagreeable surprises and serious losses.

We leave this subject, referring all those merely in-

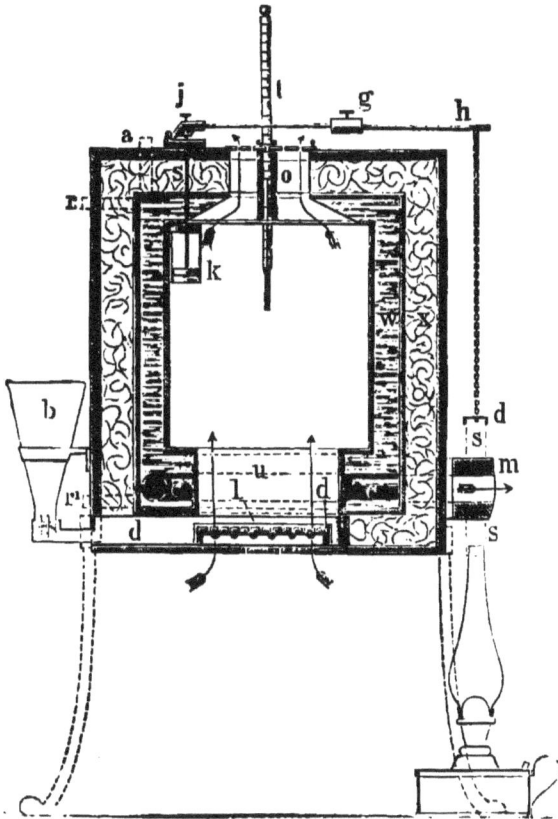

Fig. 9—WORKING PARTS OF THERMOSTAT.

terested in sterilization of milk to the treatise written by Monrad on "Pasteurization and Milk Preservation," where a synopsis of such apparatus is given, and to the article by E. A. de Schweinitz in the year book of the U. S. Department of Agriculture for 1894.

CHAPTER IX.

The Mortality of Infants.

Cow's milk is pure only in the upper part of the healthy animal's udder—the lower parts of the milk, principally that contained in the milk cisterns adjoining the teats, are, as has been previously shown, more or less polluted by germs that have found their way through the ducts in the teats. Impure milk may be, however, milk physically decomposed by distemper in the cow or by the admixture of filth, dust, hair, scales from the outer skin of the udder, germs of lower organisms, or by all these conditions combined. Watchfulness as to the sanitary condition of the cow and the observation of a scrupulous cleanliness in every handling of the milk tend to lessen the evil influences just named. It is an easy matter for every farmer or dairyman to convince himself, by a simple experiment, of the great difference in keeping qualities that result from improved conditions whilst milking.

Let him enter his stable at a given morning and milk three cows into the milk pail he has been using all along and without any change of accustomed conditions; let him mix this milk and take out a test sample for setting; let him then take the next three cows, lead them out into the open air, wash the udder, if soiled, with warm water, and dry thoroughly with a clean towel, or, if not soiled, rub gently, but thoroughly, with a moist towel, so that all dust, hair and scales may cling to it, then wash hands in water and soda, dry them, and milk into a new milk pail which has previously been well sterilized by boiling water and soda, and letting the first five strippings from every teat run to the ground, then mix the milk of these three cows by itself, as of the lot before, and place the test sample by the side of the first in the same place of storage, at a temperature of 60° or less, and he will remark that the first sample to " turn " will be the one of the stable-milked cows, which will take place in about from twenty to twenty-five hours, whilst the sample from the second lot, the one produced under improved conditions of cleanliness, will keep sweet for from ten to fifteen hours longer than the first.

After improving the conditions of milking, we may turn our attention to the straining ; and here, it must be confessed with regret, we find, in general, a sorry condition of affairs. By far too many farmers do not catch the meaning of the idea to be conveyed when speaking of microscopical minuteness. They believe that dirt, to be perceptible, must be visible, and the double or trebly-folded cloth in the strainer is considered quite an extra concession to cleanliness and fancifulness ; yet, minute particles of dirt do pass, detectable in the aggregate even without the use of the microscope as a horrifying mass of filth.

A very simple experiment may be made to convince

as of the quantity of dirt remaining in the average
stable-strained milk.   Take a clean glass vessel, of
the shape shown at Fig. 24, and containing about one
gallon of the fresh strained milk, fasten six inches of
rubber tube over the mouth of the bottle and a small

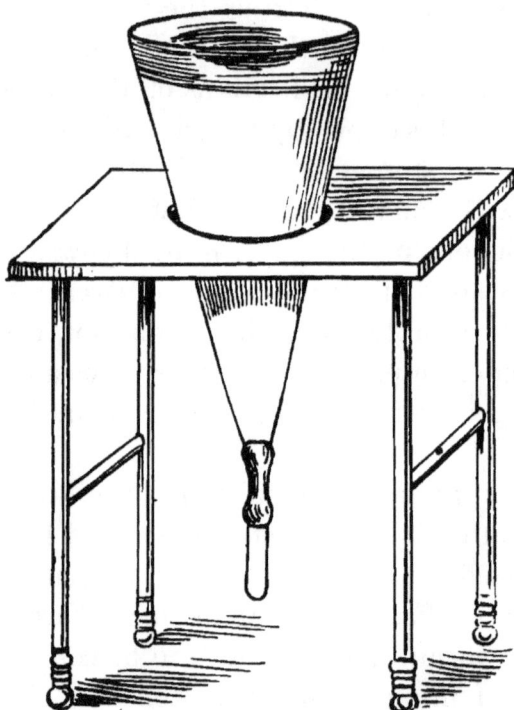

Fig. 24—TESTING FILTH IN MILK.

glass test tube to the other end of the rubber tube,
turn upside down, place in a suitable rack and let it
remain standing for twelve hours.   The dirt contained in that milk has now settled down to the
bottom of the small glass tube; this is removed by
tightly closing the rubber tube with thumb and index

finger, turning the large vessel right side up and pulling away the rubber tube from its mouth. The contents of the smaller tube are now poured over a blotting paper filter from which, after drying, the actual amount of dirt in the milk may be ascertained by weight. In this manner the percentage of dirt in the daily milk brought to market was ascertained for all the larger cities in Germany, and, as a result, figures were published that shocked the public and were pronounced incredible exagerations, until a leading scientist in dairying technics undertook to convince the public by exhibiting these dirt accumulators in operation at fair grounds and at all suitable occasions.

The majority of milk consumers in cities when bestowing a thought on the origin of the milk brought to their home by the trim milk wagon, picture the farm dairy as a scene of rural bliss and healthful surroundings, where clean glossy cows browsing in the sunshine on flowery pastures, or peacefully lying down, chewing the cud in the shade of lovely trees, have all the care and attention their importance merits.

Against this fair picture, let us hold up reality in the form of an abstract from the able report of Dr. Howard Carter, milk inspector of the city of St. Louis, Mo., for 1895–'96, covering 436 dairies with 9,000 cows: " The sanitary condition of a majority (of dairies) however, is vicious in the extreme, and their presence in the thickly populated district should not be tolerated. Deprivation of natural food, light, air, exercise

and natural environment can result only in impaired
health, whether in man or animal. There are 322
dairies having no pastures, 126 having neither pasture
nor cow lot, 77 having improper facilities for cooling
and storing milk, or none at all. The breathing
space is entirely insufficient. The majority of dairies
are badly ventilated and poorly lighted, being more
or less entirely destitute of sunshine; in not a few
there is almost complete and perpetual darkness. In
some instances the food for the cows is boiled within
the stables—the atmosphere of which is rendered still
more oppressive by the steam and smell arising from
the boiling mash; these, added to the ammoniacal odor
of decomposing urine, produce an insufferable atmosphere. Of the milk producing properties of such
food as brewers' grains and the waste products of dis-
tilleries and vinegar factories, there appears to be but
little doubt, yet authorities who have more thoroughly
investigated the subject assert that the quality of
milk produced under such feeding is less stable in its
constituents, the fat more readily broken up into the
various fatty acids, the casein less soluble and the
whole product more liable to the various forms of decomposition than milk produced from healthy animals
under natural environments. But the result of such
a system of stabling and feeding is, however, a per-
version of the natural appetites and functions of the
animals subject to them. This is exemplified in the
refusal of such animals to drink water even in hot
weather. The continued use of partially fermented

moist food producing an analogous condition to that of
chronic alcoholism in human beings. Such conditions inevitably result in diminished vitality and a
greater susceptability to disease, although our local
dairymen profess a different opinion.

"There exists a lamentable and disgraceful disregard
for the cleanliness of the cows themselves. The animals are, for the most part, confined in stalls and de-
prived of bedding, standing out their wretched lives
upon hard board floors ; they lie down in their own
evacuations, which adhere to the flanks and udders
in dense masses. Under these conditions the production of a pure milk supply is impossible. Milk thus
collected unavoidably contains impurities of all kinds,
consisting chiefly of stable litter, manure, epithelial
scales from the teats of the cow as well as from the
hands of the milkers."

The report goes on to say that about seventy-five
per cent. of the cows in these dairies were found to
be affected with tuberculosis, and the doctor urges
the necessity of bestowing a greater share of public
and legislative attention than heretofore on this matter, being one of vital importance.

It is simply wonderful what the public will stand
in the way of filthy milk, as far as this is an established fact for the various large cities in Germany, and,
if we may consider the frequent complaints found in
the various agricultural and dairying periodicals of
this country as an indication in this direction, it must

be conjectured that the state of things in America is hardly better, if not worse.

According to the most favorable calculations it was found that the inhabitants of the city of Berlin consumed, annually, in their milk, no less than one hundred thousand pounds of cow dung, and the inhabitants of the city of New York will consume at least three times this amount per year. This is the first point to be remedied. When we consider that the new-born babe consumes only milk, and that a majority of the ailments that are liable to befall it take their origin in the stomach, we must come to the conclusion that impurity of the milk must frequently be the cause.

The death rate of infants is appaling. On an average, twenty per cent. of all children born die during the first year of their life, and, out of every hundred infants that die, eighty at least have been fed on cows' milk. But even the healthfulness of mothers' milk is entirely dependant on the physical condition of the mother. Statistical investigation has shown that while of one thousand infants nursed by mothers belonging to the wealthy aristocratic classes only 57 would die ; the mothers of the poorer classes would lose 357 out of every thousand of their infants in the same time and period of life, and even this terrific loss does not tell the whole story, as large numbers of those surviving drag an impaired constitution through life, owing to the deleterious effects of the damaged and poor milk imbibed during infancy.

But mothers that nurse their own infants have, for one reason and another become very scarce, so that there is not one class of society in which natural nursing is not on a steady decline, and it is not exclusively the aristocrat that shirks this duty or the woman that has to gain her livlihood in the factory, but it is just the same with the population in the country. I have lived for nine years near a German village of over two hundred souls, and, on careful investigation, I was unable to hear of one single case during that entire period where a mother had given her infant the breast. The hiring of the services of a wet nurse is beyond the means of most mothers and even those that do resort to this expedient generally find the nurse the terror of the household.

Boiled milk is generally considered a proper food for infants, and people have thought that to boil milk at home and dilute it with water was all that had to be done to ensure a faultless article of food for the infant. A number of receipes have, in the course of time, been brought forward and tried, such as peptonizing the cow casein by the admixture of pancreas ferment or the addition of preparations of white of egg, not one of these compounds has, however, been able to receive the support of medical science, and very justly so. Simple, but not always effective, apparatus—like the *Soxhlet*—have been invented for sterilizing infants' milk at a small cost in every household, yet their utility is, in a great measure, dependent on what the quality and condition of the

7

milk has been *before* it reached the house. We know now positively that all germs contained in fresh milk: baccillæ, spores and ferments begin to multiply immediately after being drawn from the cow with an astonishing rapidity, so that milk produced under the most favorable conditions may contain millions of germs if several hours have elapsed between the drawing from the cow and the boiling or sterilizing of it. And even if we could remedy this defect by keeping a cow in every household, we should not be producing an infants' food that could be pronounced a fit substitute for the mother's breast, for we must ever remember that *cow's milk is not mother's milk, and that the new-born babe does not possess the stomach of a calf.*

Let us look at a comparison of the two milks taken from one hundred and fifty analyses:

	Cow's Milk.	Woman's Milk.
Water	87.5 per cent.	88.25 per cent.
Casein	3.0 "	0.75 "
Albumen	0.5 "	1.00 "
Fat	3.5 "	3.50 "
Milk sugar	4·8 "	6.25 "
Ashes	0.7 "	0.25 "
	100.0 "	100.0 "

We remark at a glance the great difference of proportions in the various constituents of the two milks, and when we consider that an infant's stomach is an exceeding dainty apparatus, it will be at once clear

that these differences may be the cause of grave derangements, and this, in fact, is the case.

The principal difference, and the one which before
all others claims correction, is the excess of casein in
cow's milk in a form not of easy digestion; further-
more, the scantness of milk sugar and of albumen.

Medical authorities do not seem to entirely agree
on the equality of the chemical composition of the
casein in cow's milk and in human milk; we may,
however, without attempting to express an opinion
on this matter, fix our attention on the difference in
digestability of the two caseins, as this is of prime
importance in the process of the infant's nourishment.
If a small quantity of woman's milk be taken and a
few drops of extract of rennet added, in imitation of
the process inacted in the infant's stomach, it will be
seen that this milk coagulates in the form of finest
flakes, looking more like very minute grits, while, if
we repeat this experiment with cow's milk, we shall
see the casein formed into large, more or less compact, lumps. The digesting juices of the infant's
stomach are able easily to reduce the finely curdled
casein of mother's milk, but the lumps of the cow
casein are not easily digested, cause inconvenience,
and are, as we all have had occasion to observe, frequently ejected from the infant's stomach. To reduce
the amount of casein in cow's milk by diluting with
water is a proceeding adopted by many; it is not,
however, a recipe to bring the milk any closer in
composition to mother's milk, as, by so doing, we re-

duce yet further the already deficient percentage of milk sugar, albumen and fat, the latter, especially, furnishing the greater part of strength in the infant's food, and it is exactly this strength which is so important a matter to be kept up.

Our aim in preparing a reliable substitute for the mothers' breast, in producing an artificial mothers' milk, must then be to convert cow's milk, by an absolutely harmless proceeding, into a thoroughly healthy milk, containing exactly and constantly a uniform percentage of ingredients closely resembling those contained in healthy mothers' milk and to change the form of curdling of the casein into the one proper to human milk. Simple as this undertaking may seem to a mind that has not had an opportunity to study the intricacies of the matter, this desideratum has been the life aim of many a scientist, and it is only the last few years that have brought us closer to the attainment of this boon, by the labors and successes of Prof. Backhaus, of Gœttingen, of Prof. Gærtner, of Vienna, and others, in whose mothods of converting cow's milk into artificial mothers' milk, we now possess admirably planned processes, in which every change and manipulation is founded and supported by universally accepted medical principles. The satisfaction with which this milk has been hailed by the medical men in Europe, has created a demand for it beyond all expectations, and in a very short time every city and town will possess a dairy manufacturing this artificial mothers' milk, and to judge from the numerous inquiries that have been sent from America, and from the hearty encouragement I have received from the medical men of this country, it would seem that this article will, also here, be gladly hailed, and fill the place of a true blessing. It will not be found amiss to append two testimonials from German physicians :

Dr. (med.) Hess, says : " During the epidemic of cholera infantum, in the summer of 1895, I had the opportunity of becoming acquainted with the nutritive and curative properties of the normal infants' milk. I treated eighty-two infants, part of them purely medicinally, and part of them purely dietically, another part with combined treatment, according to the age of the infants and the intelligence of the parents. On the whole, I am able to record great success in all cases where the nursing was properly attended to, where the milk was administered according to instructions, and where the infant received the milk direct from the bottle. I had eight cases of death, two of these were infants that had received the normal milk. Out of my eighty-two little patients, fifty-five were treated with the normal milk alone, fifteen received medicines besides, twelve were treated with medicines only, and of these latter, six died. The medicines prescribed were : Kreosot, argent. nitric. colombo and Bism. subnitr, according as conditions required, also Tokay wine. My opinion is, that if I were placed before the alternative to combat a case of cholera infantum, or of summer diarrhea with either the normal infants' milk, or with medicines, I should unhesitatingly try it with the first, because I have become convinced of the uselessness of the medicines without regulating the diet."

Dr. (med.) Marx, says: " During the summer of 1895 I experimented with the normal infants' milk on a number of sick and of healthly infants, reaching surprising results. In cases of summer diarrhea and cholera infantum, even where the Soxhlet milk had been given without avail, an immediate improvement followed the taking of the normal milk, vomiting and discharges ceased, giving place to a healthy digestion. In healthy infants, where nursing by the mother was impossible, and the normal milk given, I found an average daily increase in weight of 30 grammes during the first months of life. Cases where the normal infants' milk did not agree at all, or even where it did not well agree with an infant, have not come under my observation."

Professor *Escherich* says : " It is a well known fact that, even with the aid of the most perfect hygienic conditions, infants with satisfactory digestion, but not brought up on the breast, do not show the same resistancy against sickness that breast-infants do. It is to be hoped that by the introduction of the normal infants' milk the percentage of failures will be lessened. The normal milk may be given to infants of all ages, but is more particularly indicated when infants, for some cause or other, take too little food, and which, in consequence of insufficient nourishment and intercurrent ailings, have been stunted in development, also to infants which are to be weaned from the breast, or where the breast is not entirely sufficient, and to such which possess particularly irritable organs of digestion. The pugnacious constipation so often noted in infants that take diluted or undiluted cow's milk will vanish with normal milk and reappear when changed back to the former. Only in those forms of acute indigestion that end with diarrhea, and in which milk in any form is not supported, also the administration of normal milk should be suspended and another regime prescribed by the physician. In all other chronic forms of indigestion and indications of weakness a heightened assimilation of fat is of importance, as this factor of nourishment is particularly well absorbed by the infantile colon without any precursory enzymotic transformation. Clinical observations have been made in this direction by *Biedert, Banze, Demme* and at *Monti's* Polyclinic. The great advantage which normal infants' milk posesses, as compared with other " prepared " or " modified " milks, is that it contains a proper percentage of fat but only a third part of the casein, which is so difficult of digestion, and it is just this fat which allows of a copious supply of calorics without overburdening the digestive organs. An idea prevails that younger infants require a nourishment of different composition than older ones and that mothers' milk undergoes a change with the advancing age of the infant. The more recent investigations have, however, refuted this assumption. It has been found that, apart from the first fortnight, the milk from one and the same wet-nurse did not materially change during the entire nursing period. *Soxhlet, Heubner* and others recommend to follow the example set by nature and to prepare the normal milk to one unvarying standard, and experience has proved this to be correct. A most valuable feature is the steady increase in weight of infants that take normal milk. Professor Escherich has published the results of his investigations in this line; from them I take one example :

Week of Life.	Weight of Infants, in Grammes.	Weekly Advance, in Grammes.	Quantity of Normal Milk Taken.
23d	5,675	——	1,300
24th	6,000	325	1,300
25th	6,500	500	1,300
26th	6,775	275	1,750
27th	6,900	125	1,750
28th	7,100	200	2,000
29th	7,350	250	2,000
30th	7,575	225	2,000

This infant, when receiving normal milk for the first time, weighed 5,675 grammes, while the normal weight of a babe twenty-three weeks old has been found, by *Camerer*, to average 6,132 grammes. The infant was, therefore, lighter by 457 grammes than a normal infant. Now, the average advance in weight of an infant between the twenty-third and thirtieth week has been ascertained at 719 grammes, for such

as are nursed on the breast, and 818 grammes for those
artificially nursed. The infant in question had, however, made a gain of full 1,900 grammes, and at the
end of the period of observation was 625 grammes
heavier than a normal infant, it had, in other words.
caught up its deficiency and made a big advance.
Another striking example is given of a younger
infant, a baby girl, in the Gras hospital :

Week of Life.	Weight of Infant, in Grammes.	Weekly Advance, in Grammes.	Quantity of Normal Milk Taken.
3d	3,600	——	800
4th	3,850	250	900
5th . . .	4,175	325	1,000
6th . . .	4,400	225	1,000
7th . .	4,650	250	1,200
8th . .	4,800	150	1,300
9th . . .	5,160	360	1,300
10th . . .	5,150	10	1,200
11th . . .	5,280	130	1,240

In eight weeks this infant had gained 1,680 gr.,
while infants artificially nursed and of the same age
only average a gain of 1,109 gr., and children on the
breast 1,582 gr. ; we must here take into consideration that the hospital is no ideal field for experiments
in rearing infants on the bottle.

The transit from common milk to normal milk is,
generally, accompanied by the immediate cessation of
any abnormal activity of digestion ; it will be well,
however, in all cases, to proceed cautiously. Dr.
Steiner remarks in his report on experiences with

normal infants' milk : " Dyspetic infants I give a day
of fasting, that is, they are put on Russian tea—*ad
libitum*—and commence the treatment with calomel
or an irrigation. I have never ventured to pass from
the dyspepsia-producing food to the normal milk
without this pause of twenty-four or thirty-six hours
and without cleansing the digestive tract. In chronic
dyspepsias I commence with an irrigation and follow
up, partly with acid. muriat. dilut. 0.7–1.0 : 200 one
teaspoonful every two hours, or magist. bismuthi 1.0
–2.0 : 100, or tinct. rhei 1.0 : 100.c. Where there is
inclination to vomit I give the milk cold. Scrupulous cleanliness of feeding bottle ; feeding nipple to be
put on milk bottle direct. Punctuality in giving the
meals and in the pauses that have been fixed upon.
For the normal milk I have found as the best intervals—cases of premature birth excepted :

For the first week	$2\frac{1}{2}$ to 3 hours
First to second month . . .	3 hours
Third to fifth month	$3\frac{1}{2}$ hours
Sixth to twelfth month . . .	4 hours

"During the night one or two feedings. From the
tenth month onward other food in connection with
the milk. If infants find the intervals too long, I
give boiled, and subsequently cooled, spring water
with a spoon. The strict observance of the quantities of milk given has proved to be less urgent than
the strict observance of the intervals. On the whole,
I have found the quantities given in the following

table sufficient, although the requirement changes with the individual. With weak infants, and such that are reconvalescent from Dyspepsia, I always prescribe the I. grade of normal milk in somewhat smaller doses, augmenting them gradually :

Age of Infant.	Feeding Interval, hours.	Single dose, gr.	Number of meals in 24 hours.	Quantity consumed daily, gr.
1 week	2½–3	30– 50	7–8	250– 300
1 month	3	50–100	7	350– 700
2 months	3	100–150	7	700–1,050
3 months	3½	100–150	7	700–1,050
4 months	3½	150–200	6–7	900–1,400
5 months	3½	150–200	6–7	900–1,400
6 months·	4	150–200	6–7	1,000–1,400
7–9 months	4	250	6	1,500

"After dyspepsia I have found the recuperation of weight even more rapid than in breast infants."

Many believe that two kinds of milk are injurious to an infant. This is erroneous. Normal milk can be given with greatest advantage together with mothers' or nurses' milk ; it should naturally be of faultless quality, and adapted to the digestive forces of the infant. Professor Gærtner, of Vienna, gives the following experience with the feeding of twin babies who, together, possessed but one nurse, and a very poor one at that. From the fifth week of their lives, onward, they received, each, about a pint of the normal milk daily, their gain in weight may be seen from the following table :

CHARLOTTE F.			MELANIE F.		
Week of Life.	Weight.	Gain, in Grammes.	Week of Life.	Weight.	Gain, in Grammes.
5th .	3,500	——	5th .	3,250	——
6th .	3,800	300	6th .	3,500	250
7th .	4,350	550	7th .	3,930	430
8th .	4,600	250	8th .	4,250	320
9th .	5,000	400	9th .	4,630	380
10th .	5,250	250	10th .	4,850	220
Gain in 5 weeks, 1,750 gr.			Gain in 5 weeks, 1,600 gr.		

These infants were a picture of health, and never showed the slightest inconvenience in consequence of their variegated bill of fare.

The success of these investigations led to others in the direction of ascertaining the effects of normal milk on adults. In complaints of the stomach, as well as in other derangements, for instance, those accompanied by fever, the activity of this organ is seriously depressed. The segregation of gastric juice is insufficient, or even entirely paralyzed ; the food eaten is not digested in a certain space of time, but remains for a longer period, passes to fermentation and decomposition, engendering the well known symptoms of serious indigestion. A nourishment which exacts no strain on the digestive forces of the stomach should be offered to such patients. We know that the mere physical function of the stomach is to transform the food eaten into a homeogenous slop.

The investigations of v. Mehring have shown that fluids are not assimilated in the stomach. Every drop

of wine, water or beer we consume passes to the
colon, which is the true organ of resorption. When
we compare the immense quantities of fluids some
people are able to absorb, with the limited capacity
of the stomach, we may conjecture that these liquids
do not remain in the stomach for a very long time,
and that they cannot be subjected to digestion in the
stomach. This is the explanation why, in serious
derangement of the functions of the stomach, liquid
nourishment alone is supported. When speaking of
liquid nourishment we are apt to think of broth and
milk.

Now, it is known that beef-broth is rather an in-
centive a stimulant than a nourishment, and that we
should never succeed in keeping a person alive on
broth alone, while milk contains every ingredient necessary to the building up and sustenance of the organism. Is milk, however, a *liquid* nourishment?
It is so only as long as it is outside of the stomach.
On arrival in the stomach it is curdled, transformed
into a lump by the acid and the rennet present, and·
this lump must be dissolved again by the gastric
juice. Bearing this in mind, we must call cow's milk
a solid food, and not a liquid one. Physicians find
this corroborated in their daily practice. Here is the
all important difference between woman's milk and
cow's milk, for woman's milk remains liquid, or, what
is the same, curdles in so minutely fine flakes in the
stomach that it is able to pass on from it without previous digestion.

We have proof that this principle has been known
and made use of in antiquity, hundreds of years
before the advent of Christ. The physicians, Eury-
phon and Herodikes, living at the time of Hippocrates
(460 to 387 B. C.) had published a method of curing
dyspepsia, making their patients take woman's milk
from the breast, direct. If we are, therefore, able to
manufacture normal milk in exact imitation of
mothers' milk, then, we produce a liquid nourishment which does not remain in the stomach but a
very short time, and does not put any strain on its
functions. Buttermilk and whey have the same property, only they are deficient in principles of nourishment. A special indication for normal milk is to
diabetics ; the milk is then specially prepared without the addition of milk sugar. Most successful
treatments are on record with this classs of patients,
thousands of whom are taking the normal milk
regularly, up to three liters per day.

CHAPTER X.

Artificial Mothers' Milk—Normal Infants' Milk.

From what has been said in the preceding pages, we become aware that the end to be attained, is the transformation of pure cow's milk into a milk, which, in its nutritive elements, is analogous to mothers' milk, the composition of which is of a constant uniformity, and its keeping qualities allow of its being trans- ported to great distances, and undergo all changes of temperature experienced during summer transporta- tion for a lengthened period, without spoiling or any way changing. We have also seen that the first step to be taken in this direction is the supervision of the production of the raw material, the exaction of scrupulous cleanliness in the keeping of the milk cows and the utensils employed, as well as an unre- mittant control of all conditions influencing the phy- sical welfare of the cows, and of the quality of the food fed to them.

In a subsequent chapter will be laid down what should be exacted to insure a healthy condition of the milk. We now pass on to the manufacture of this milk into artificial mothers' milk—normal in- fants' milk—in two grades, the first to resemble mothers' milk in the exact proportion of all nutritious ingredients, and to be a perfect and wholesome substitute for mothers' milk for infants from the time of birth up to the fourth month; the second grade of normal milk to contain the same percentage of fat, albuminoids and milk-sugar, but having a slightly higher percentage of casein, being intended to be given to infants after the third or fourth month of their lives, and to form a transitory food from the first grade of milk to pure cow's milk, a most necessary precaution, when we take into account the extreme difficulty experienced by the infant stomach to digest the casein in pure cow's milk.

In undertaking to describe the various operations destined to transform cow's milk into normal infants' milk it must, right here, be admitted that no description, however lucid, will enable a beginner to produce the desired article from the start, there being connected with the whole proceeding a number of small manipulations and advantages, which although in no manner business secrets (as some would try to make them out and guard them from the public) yet are proceedings which are only mastered by practical experience and personal application. In Germany, Austria and France, where the manufacture of normal infants' milk is rapidly gaining ground, this apparent difficulty is by no means considered a disadvantage, but, quite the contrary, as a protection, as it tends to keep at a distance that class of competition which would speedily tend to discredit normal milk.

The principal operations we shall have to follow will be:

The testing of the cow's milk for fat percentage and acidity.

The separating into cream and skim milk.

The reduction of the casein in the skim milk and the transformation of the remaining into the finely coagulating form.

The mixing, sugaring and bottling.

The sterilizing and the testing of the sterilized milk as to its keeping qualities and its freeness from germs.

Starting on the assumption that the manufacture is to be connected with an established dairy, and, as we shall see later on, the manufacture of the normal milk and the maintenance of the dairy, is inseparable one from the other if any guarantee of purity is to be attained it will then generally be found advantageous for the beginner to pass the milk over a system of coolers immediately after drawing, and this will become an absolute necessity where the evening's milking has to be turned into normal milk on the following morning.

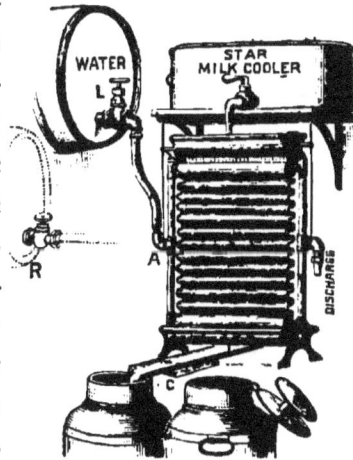

Fig. 23—STAR MILK COOLER.

The milk, as it runs from the cooler, is collected in

к

large receiving vats, where it may be thoroughly
mixed. The first proceeding is to make sure of the
percentage of fat contained in the entire quantity of
milk. If the same cows are milked daily for the
manufacture of the normal milk, and the same food
fed to them without change, then it will suffice to
take the fat test but once a week ; if, however, a new
cow has been brought in, or one of the old ones dis-
charged, or the feed been changed in any way, then
a test will be necessary as often as one of the indicated changes has occurred. To take a fair test
sample, the milk should previously be well stirred
with a wooden paddle for two minutes consecutively.
There are milk samplers, like the *Scoville*, in the
market, yet a common white glass tube, three-eighths
of an inch inside diameter, will answer the purpose
equally well. Its length should exceed by six inches,
more or less, the depth of vessel in which the milk is
contained. This tube is dipped into the milk, the
upper end closed by pressing on the thumb. When
the tube has reached the bottom of the vessel, the
thumb is removed, the lips are applied, and, by a
steady suction, drawing the tube upwards out of the
milk slowly, the tube is filled with milk from all
parts of the vessel. This is repeated three or four
times, emptying the samples into a glass dish. If the
milk to be turned into normal milk has been collected
in several different vessels, then the test samples have
to be taken from each and every one, and in a fair
proportion to the contents of each vessel, so that if,

for instance, four glass tubefuls have been drawn
from a vessel containing forty quarts, then from a
vessel containing but thirty quarts only three tubefuls should be drawn, or two from another vessel
containing only twenty quarts, etc. The test samples
are all collected in the same dish and the testing at
once performed. The temperature of milk for testing
should be 62° F., more or less. Two colateral tests
should be made of every sample to avoid errors.
Quite a number of methods and apparatus for testing
have been invented, the most accurate being probably
the *Soxhlet;* for use in dairies, however, this method
is too complicated, and the best known tester in this
country is the Babcock.

According to the instructions kindly furnished me
by Prof. S. M. Babcock, of the Wisconsin Agricultural
Experiment Station, the method of operating the test
is as follows :

THE BABCOCK TEST.

The estimation of fat in milk by this test is accomplished by adding to a definite quantity of milk,
in a graduated test bottle, an equal volume of com-
mercial sulphuric acid of a spgr. of 1.82–1.83. This
acid dissolves the casein, setting free the fat, which is
then completely separated from the liquid in the
bottle by whirling in a centrifugal machine. Hot
water is afterwards filled into the bottles to bring the
separated fat into the graduated neck, where the per
cent. is read directly from the scale.

MAKING THE TEST.

Sampling the Milk.—Accurate tests can only be obtained when the cream is evenly distributed throughout the whole mass of milk. This is best accomplished by pouring the milk a number of times from one vessel to another. Pouring three or four times will be sufficient for fresh milk fresh from the cow. Milk that has stood until a layer of cream has formed, should be poured more times, until all clots of cream are broken up and the whole appears homogenous.

MEASURING THE MILK.

When the milk has been sufficiently mixed, the milk pipette is filled by placing its lower end into the milk and sucking at the upper end until the milk rises above the mark on the stem; then remove the pipette from the mouth and quickly close the tube at the upper end by firmly pressing the end of the index finger upon it

Fig. 10.
MILK
PIPETTE

to prevent access of air. So long as this is done the milk cannot flow from the pipette. Holding the pipette in a perpendicular position, with the mark on the level with the eye, carefully relieve the pressure on the finger so as to admit air slowly to the space above the milk. In order to more easily control the access of air, the finger and end of the pipette should be dry. When the upper surface of the milk coincides with the mark upon the stem, the pressure hould be renewed to stop the flow of milk. Next

place the pipette in the mouth of one of the test
bottles, held in a slightly inclined position so that the
milk will flow down the side of the tube, leaving a
space for the air to escape without clogging the neck,
and remove the finger, allowing the milk to flow into
the bottle. After waiting a short time for the pipette
to drain, blow into the upper end to expel the milk
held by capillary attraction in the point. If the
pipette is not dry when used, it should be filled with
the milk to be tested, and this thrown away before
taking the test sample. If several samples of
the same milk are taken for comparison, the
milk should be poured once from one vessel
to another before each sample is measured.

ADDING THE ACID.

Great care should be taken in handling the
acid, as it is very corrosive, causing sores upon
the skin and destroying clothing unless quickly removed. If, by accident, any is spilled upon the
clothes or hands, it should be washed off immediately, using plenty of water. A prompt application
of ammonia water to clothing upon which acid is
spilled may prevent the destruction of the fabric, or
restore the color.

Fig. 12.
ACID
MEASURE.

The acid measure is filled to the mark with sulphuric acid and carefully poured into the test bottle
containing the milk to be tested. This bottle should
be held in a slightly inclined position, so as to allow
the acid to run down the side of the bottle. The

acid is heavier than the milk and sinks directly to the bottom, forming a clear layer. The acid and milk should be thoroughly mixed together by shaking at first with a rotary motion until the curd which forms is entirely dissolved, and then completed with a vigorous shake sideways. A large amount of heat is evolved by the chemical action, and the liquid changes gradually to a dark brown.

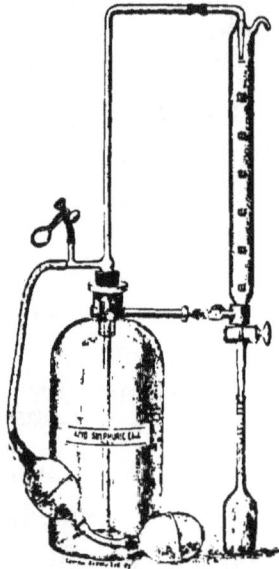

Fig. 11—IMPROVED ACID BURETTE.

WHIRLING THE BOTTLES.

The test bottles containing the mixture of milk and acid should be placed in the machine directly after the acid is added. An even number of bottles should be whirled at the same time, and they should be placed in the wheel in pairs opposite to each other, so that the equilibrium of the apparatus will not be disturbed. When all the test bottles are placed in the apparatus, the cover is placed upon the jacket, and the machine turned at the proper speed for about five minutes. The test should never be made without the cover being placed upon the jacket, as this not only prevents the cooling of the bottles when they are whirled, but, in case of the breakage of bottles, may protect the face and eyes of the operator from injury by pieces of glass or hot acid.

FILLING THE BOTTLES WITH HOT WATER.

After the bottles have been whirled, they should be filled immediately, with boiling water, to the neck, and then whirled again for about one minute, and more water added to bring the fat into the graduated neck. A third whirl of about one minute is given to bring all of the fat into the neck where it can be measured.

MEASURING THE FAT.

The fat should be measured immediately after the whirling is completed, before it has cooled to a point where it does not flow freely. If many tests are to be made at the same time, better results are obtained by setting the bottles in hot water to keep the fat in liquid condition until the readings can be taken. To measure the fat, hold the bottle in a perpendicular position with the scale on a level with the eye, and observe the divisions which mark the highest and the lowest limits of the fat. The difference between these gives the per cent. of fat directly. The readings should be taken to half divisions of the scale, or to one-tenth per cent.

The readings may be made with less liability of error by measuring the length of the column of fat with a pair of dividers, one point of which is placed at the bottom and the other at the upper limit of the

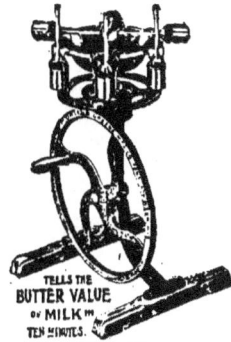

Fig. 13.
SMALL WHIRLING MACHINE.

fat. The dividers are then removed, and one point placed at the 0 mark of the scale on the bottle used, the other will be at the per cent. of fat in the milk examined.

Skim milk, buttermilk and whey are tested in the same general manner as full milk, except that skim milk and buttermilk require about one-fourth more acid and should be whirled about two minutes longer than whole milk, while whey requires only about two-thirds as much acid as milk. Where the amount of

Fig. 14—STEAM TURBINE WHIRLING MACHINE.

fat is less than two-tenths per cent. it often assumes a globular form instead of a uniform layer across the tube; where this occurs, the per cent. of fat must be estimated. In doing this, it must be remembered that any appearance of fat in the tube indicates as much as .05 per cent. It is not possible, with the Babcock test, to detect less than .05 per cent. of fat.

CREAM.

Special bottles are provided for testing cream. The operation is the same as with milk, except that the

cream adhering to the pipette should be rinsed into
the bottle with a little water, and, after the acid is
added, the bottle should be allowed to stand for
about five minutes before it is whirled. During this
time it should be shaken occasionally, and if the room
is cold the bottle should be kept hot by setting in hot
water.

Cream may be tested in the ordinary bottles by di-
viding the test sample, as nearly as can be judged by
the eye, into three bottles. The pipette is then rinsed
twice into the three bottles with water, and the test
made as with milk, the readings upon the three bottles being added together for the per cent. of fat.

Where a balance is available, the best method is to
weigh the cream into an ordinary test bottle, taking
about five grammes for a test, and adding to this
about 12 c. c. of water. The test is then made as
with milk, the readings being multiplied by eighteen
and the product divided by the number of grammes
of cream taken for the per cent. of fat.

Condensed milk is tested in the same manner as
cream. The sample should always be weighed, as
these milks are usually too thick to be accurately
measured with a pipette.

As we may surmise, the fat test is one of greatest
importance towards insuring an unvarying quality in
the normal milk. The result of the tests should be
kept on record, as they are of value to indicate the
influence which changes in the feed have on the percentage of fat in the milk.

Besides the fat test, it becomes necessary, periodically, to make a test of the acidity of the milk to be used ; this is more particularly the case in hot weather, or where ensilage is fed, or any apprehension exists as to the sweetness of the fodder or pasturage. For the acid test, 50 cub. cent. of milk are placed in a glash dish, 2^{cc} of hydrate of sodium and two or three drops of phenolphtalein added and mixed together. To this we now cautiously add common sulphuric acid, by means of a graduated pipette, constantly stirring, until a decidedly pink tinge appears in the milk. When this has set in the accurate quantity of acid added in c. c. is ascertained, and we call every cubic centimeter added one degree of acidity. In this way milk to be used in the manufacture of normal milk may contain no more than three degrees of acidity, any excess of this quantity will tend to spoil the milk—to make it curdle. Milk that shows 4.5 degrees of acidity is unfit for the manufacture of normal milk. Milk which has turned sour shows 26.5 degrees of acidity ; butter may show 15 degrees.

If we have found our milk sweet we now proceed to the separation of the cream from the skim milk, conducting the milk into a tempering vat where it attains a temperature of 86° F. The separator is graduated in a manner to turn out one-third of the volume of the milk as cream and two-thirds as skim milk. This must strictly be adhered to, as on this division all subsequent calculations are based. After the separator gets first started, four or five gallons of the skim milk are caught in a separate vessel and put aside, to be passed through the separator again with

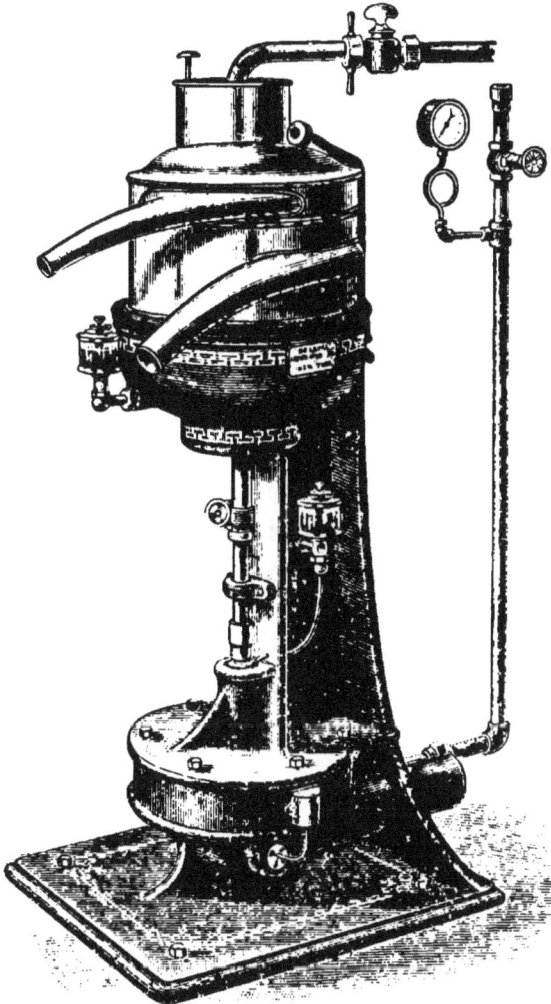

Fig. 15—De LAVAL STEAM TURBINE CREAM SEPARATOR.

the last of the milk. Any good separator may be used; where, however, larger quantities are to be produced, the use of the steam turbine separator is to be recommended, and the *de Laval* has here given universal satisfaction. When all milk has passed through the separator the scales are used to ascertain if the separation has been effected in the prescribed proportions, returning some of the skim milk to the cream if this latter had not come up to one-third of the entire quantity.

The percentage of fat to be given to the normal milk is three per cent., or one hundred pounds of milk should contain three hundred units of fat; a richer milk will therefore have to be reduced by the addition of skim milk, or by the retention of a portion of the cream ; a milk poor in fat will have to be enriched by the addition of cream, or by the retention of part of the skim milk. As an example : We wish to use 200 pounds of milk testing 4.2 per cent. of fat ; we separate this into

86.6 pds. cream —— 173.4 pds. skim milk.

As we wish our normal milk to contain but 3 per cent of fat, we must find out how much of this cream will have to be returned to the skim milk to result in a milk of the desired percentage.

$$4.2 : 3 = 86.6 : x$$

$$\text{or,} \quad \frac{3 \times 86.6}{4.2} = 61.8 \text{ pds.}$$

The reverse will be the case where milk is found to be below the required standard of fat percentage.

The cream vessel is now covered, placed in a cold
water bath and put out of the way while we proceed
to extract the excess of paracasein from the skim
milk, a process in which 15 per cent. of the original
weight of the skim milk is lost, and which is an item
to be taken into account when making calculations
for fixed quantities required. Tables of figures have
been prepared to show the quantities of cream and
skim milk with reference to the different percentage
of fat and the loss of paracasein for the preparation
of both grades of milk.

We may call to mind what has been previously said
on the simple mixtures of milk, cream, water and
milk sugar, which do good service to older infants,
when properly prepared, but are not adapted for consumption by the new-born babe; because the albumen
in them is administered, principally, in the form of
cow casein, which latter will, according to the experiences of *Biedert*, continually be accompanied by
deleterious effects, even if its form of coagulation has
been somewhat changed by the manipulation it will
go through in this process.

The more recent elementary analysis of *Wroblewsky*
seems to prove, without doubt, that a most distinct difference exists between cow casein and human casein.
If the diluting of cow's milk is carried to a point
where only one per cent. of casein is left in the milk—
the limit of quantity which the infant's stomach
will endure—then there is a deficiency of albumen and
salts. Corresponding to the large admixture of water,

we also find it necessary to give a heavy dose of milk
sugar, by which the costs of the manufacture would
be greatly enhanced. By some, it has been tried to
substitute the cheaper cane sugar, but this has proved
a failure on account of its greater propensity to turn
acid in the infant's stomach, and because milk sugar
possesses special properties of the greatest importance,
to ignore which would be equivalent to endangering
the reliability of the entire process of turning cow's
milk into artificial mothers' milk. The chemical and
physiological action of milk sugar on the organism
cannot be substituted by either maltose, glucose or
cane sugar. To imitate nature—an ever reliable
practice in similar cases—has here not proved to be
an effective argument, as milk sugar plays but an in-
significant part in the customary nourishment of infants, while the most unnatural admixtures: the
starchy matter contained in so-called infant foods, are
frequently resorted to. *Soxhlet* found the absolute
necessity of milk sugar to the infant founded on the
following differences between it and other sugars:

1. Excepting cane sugar, which for other reasons
cannot be considered, milk sugar is the only kind of
sugar which, when heated with nitric acid, produces
slimy acid, while the other sugars produce sugar acid.

2. Cane sugar, maltose and glucose disintegrate in
the presence of common alcoholic ferment into alcohol and carbonic acid; milk sugar remains un-
changed, and resists to all fermentative influences by
far more powerfully.

3. Milk sugar possesses only about one-third of the sweetness of cane sugar ; we are, therefore, able to mix three times the quantity to a nourishment without producing a repugnant sweetness.

4. It is not transformed like the other sugars into glykogen, has an enhanced combustability and passes easily into the urine.

5. Maltose and cane sugar are the most rapidly absorbed, milk sugar but very slowly ; 70 to 80 per cent. of the former in one hour, of milk sugar but 20 to 40 per cent., depending on the strength of solution.

6. The accumulation of the rapidly absorbed sugars in the blood produces very notable changes in the functions of the apparatus of circulation, which persist until the blood is relieved of this excess of sugar. The pressure of blood is heightened, the vessels become expanded, the pulse is augmented, circulation is so much accelerated that double the quantity of blood passes through the same vein during a measured span of time. Milk sugar produces quite a unique effect on the circulation ; although the blood pressure is equally enhanced if given in large doses, yet the pulse is not accelerated, but rather diminished, producing an ample systole. The heightened pressure of blood is caused by the irritating effect the other sugars have on the heart and its vessels ; the diminishing of the pulse is ascribed to the specific influence of the milk sugar on the checking apparatus of the heart.

7. While the other sugars are nearly entirely ab-

sorbed through the stomach, there will always pass
a considerable quantity of the milk sugar to the colon,
where it invariably produces a heightened secretion
of slime and gall, and by this means acts slightly
purgative. It is particularly to this specific effect of
milk sugar that attention should be drawn, as it
makes milk sugar not only an invaluable, but also a
most necessary, admixture to artificial mothers' milk.

Kehrer had conceived the idea of producing an
infants' milk by mixing the whey produced in cheese
factories with cream, but after exhaustive experiments this proved to be unsatisfactory, on account of
such whey being too poor in albuminoids, besides
being too strongly polluted with bacteria, having acquired a pronounced change in taste and commonly
possessing an amount of acidity by far in excess of
any to be tolerated in the manufacture of normal infants' milk. In a like manner it has been tried to
make use of cream procured from creameries, but
with equally unsatisfactory results, this cream being
strongly infected with bacteria, and the butter fats
so strongly influenced by improper feeding that the
palatability and keeping qualities of the normal milk
are greatly impaired. These experiments have, however, proved invaluable, by showing the way on which
the desired end might be reached.

If we treat fresh, clean cow's milk by a properly
prepared rennet ferment, observing proper temperature, time of acting, and special method of stirring,
we are able to produce an albuminous milk serum,

because this ferment has dissolved the casein into paracasein and soluble peptonic whey-protein, of which only the first named is expelled as a stiff curdled sediment.

All the albumen of the milk and all of the milk sugar are retained in this serum, and if our milk has been produced under observation of all precautions herein enumerated, it will be of an agreeable, sweetish taste and its acidity so small that the albumen—which in common whey, separates at 158° F., in consequence of the higher acidity—remains incorporated up to much higher temperatures, so that an effective sterilization is possible without damaging the nutritive qualities of the proteids. This is a delicate process, furnishing, however, a milk serum containing one per cent. of albumen, composed of easily digestible albuminoids, the whey protein and lacto protein, and, besides, five per cent. of milk sugar. If this fluid is condensed to four-fifths of its volume by the use of a vacuum pan, then we attain 1.25 per cent. of albumen and 6.25 per cent. of milk sugar. By the addition of cream we attain one-half per cent. of casein and from 3 to 3.5 per cent. of fat, a combination analogous in every respect to mothers' milk.

The percentage of ashes and salts is, undoubtedly, somewhat higher in this prepared milk than in mothers' milk, although by the action of the ferment the percentage of salts has been reduced. Normal milk shows an excess of 0.3 per cent. of salts over mothers' milk, but elaborate experiments have shown that this excess is not only harmless, but, on the contrary, entailing an augmented percentage of phosphate of lime, and therefore welcome in the systems of all infants disposed to attacks of scrofula, rachitis and kindred ailments. The ferment employed in the extraction of casein is prepared by a process exclusively adapted to laboratory work, and may, therefore, be advantageously left to those, who are by training better fitted, to attend and watch a process which requires a number of scientific appliances to produce an article of unvarying strength and composition. It is this part of the manufacture only which is not in the hands of the dairyman, but experience has shown that this is rather an advantage than otherwise. Without taking into consideration the time it would take the dairyman to produce the ferment for his own use, the production in the laboratory on a large scale can be effected with much greater economy. The properties of this ferment are:

1. That it imparts to the milk the slight alkaline reaction which we note in the woman's milk, and which, undoubtedly, must be considered as an essential factor in the process of digestion.

2. That it dissolves a part of the casein; so that we attain to an equal amount of digestible albumen, the same as in woman's milk.

3. That it curdles the paracasein and transforms the remaining casein into the form or fine flaked curdling. The strength of the ferment is continually tested and the quantity required for curdling is clearly

printed on every package. We now proceed to the
operation of curdling. The skim milk is placed in a
vat especially constructed for the purpose, fitted with
enveloping steam jacket and heated 104° F.; the
ferment is now added in the exact proportion which
the strength of the ferment calls for; the milk is

Fig. 16—CURDLING VAT.

now stirred for three minutes; the vat is then covered
and left for fifteen minutes, when the stirring is renewed with a paddle until curdling sets in, which
should take place about thirty minutes after adding
the ferment. Instantly after curdling has taken place

steam is turned into the steam jacket and the temperature brought up to 122° F., where it is kept during the time necessary to remove the lump of paracasein, which has now formed on the bottom of the vat, and which is effected by means of sieves fitting snugly into the bottom of the vat. The remaining whey will be found with agreeable, sweet taste but must not retain any sediment of casein. The vat is now heated to 167° F. and kept at this temperature for forty-five minutes to deaden the effect of any ferment remaining, great care being required not to exceed this temperature, or the albuminoids will become indigestible. At this stage of proceedings it is well to call to mind that no utensils or vessels must now be dipped into the serum, or whey, which previously have been used in fresh milk or cream. After the elapse of the forty-five minutes of heating, the serum is now returned and mixed with the cream previously separated from it, until it appears as one homogenous fluid. Where condensing is not applied to highten the percentage of milk sugar this latter must now be added (five grammes per pound), thoroughly mixed with the normal milk, which is at once bottled and ready for the sterilzing apparatus.

Before following this milk to sterilizing, we turn to the manufacture of the second grade of normal milk. The fresh milk is separated into one-third part cream and two-thirds parts skim milk, the same as for the first grade, and the calculation of fat percentage performed in the same manner. The casein in this skim milk is, however, not extracted, but only reduced by removing one-half of the entire quantity of skim milk and replacing it by pure water, with the addition of twelve grammes of milk sugar per pound of milk manufactured.

As to the advisability of using milk rich in fat, or such which is less so, will depend on the profitable use the remaining cream or skim milk can be put to. Where an equal demand exists for both grades of the normal milk, there will, when using a milk with less than 3.3 per cent. of fat, always remain a surplus of skim milk. In the manufacture of grade I. alone, there will nearly always be a surplus of cream, while in the manufacture of grade II. alone, there will always remain on hand a surplus of skim milk. As a general direction, it may, however, be laid down that milk, to be profitably used up, should not fall below three per cent. of butter fat.

If bottles of different color are not used for the I. and II. grades of the milk, then proper precaution must be provided so that bottles with different contents do not get mixed in sterilizing. Various bottling devices and apparatus are in use—a very good one is made by Boldt & Vogel, of Hamburg.

The bottle to be used is shown in Fig. 19; it is manufactured in three sizes, to contain four, seven and ten ounces each of "normal milk." As soon as filled, the rubber caps are drawn on the bottles by hands scrupulously clean.

The innumerable changes that have been brought

out in sterilizing machines, during the last few years, are, in themselves, proof of the general deficiency of these machines. I shall draw attention to the one that has given great satisfaction in sterilizing the normal infants' milk. It is built to my order by the Dairyman's Supply Co., of Philadelphia, and shown

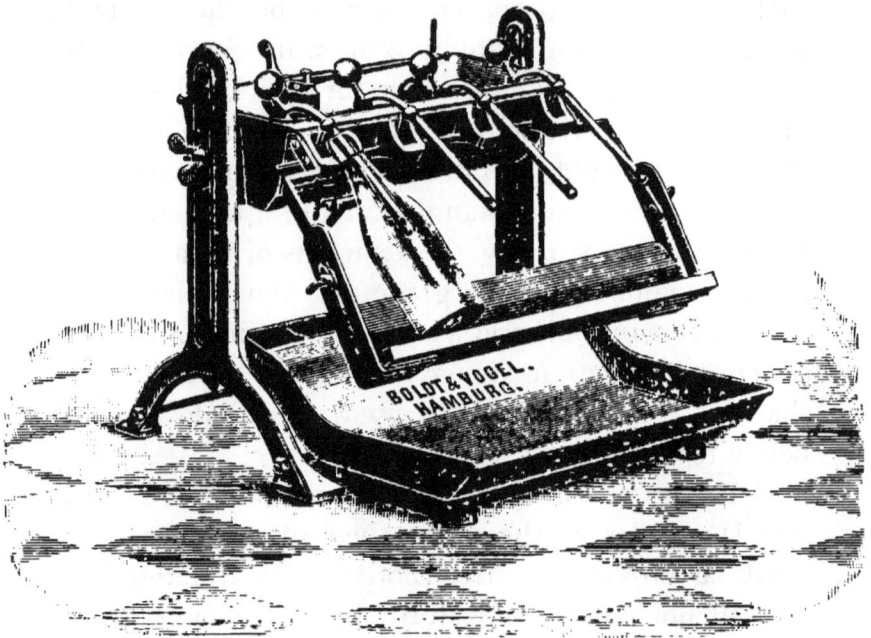

Fig. 17. AUTOMATIC BOTTLING APPARATUS.

in Fig. 18. A is the bed plate with heavy flange and rubber packing, on to which the hood or dome B is lowered and securely fastened by clamps all around. D is an upright metal tube carrying the shelves or plates C, on which the milk bottles are placed. These shelves are adjustable to different height and distance from each other to accommodate different sizes of bottles. E is a metal arm or bracket to carry the bottle, into which the thermometer dips to register the temperature of the milk in the bottles during sterilization. A second thermometer, F, is necessary to show the gradual heating of apparatus. This is a most necessary precaution, without which, considerable breakage of bottles is unavoidable. The steam enters at S, ascending by the central tube D, and passes out on to the shelves by numerous holes. Through T cold air can be forced into the apparatus, this tube connecting with the ice house. G is an exhaust pipe for carrying off the air at the beginning

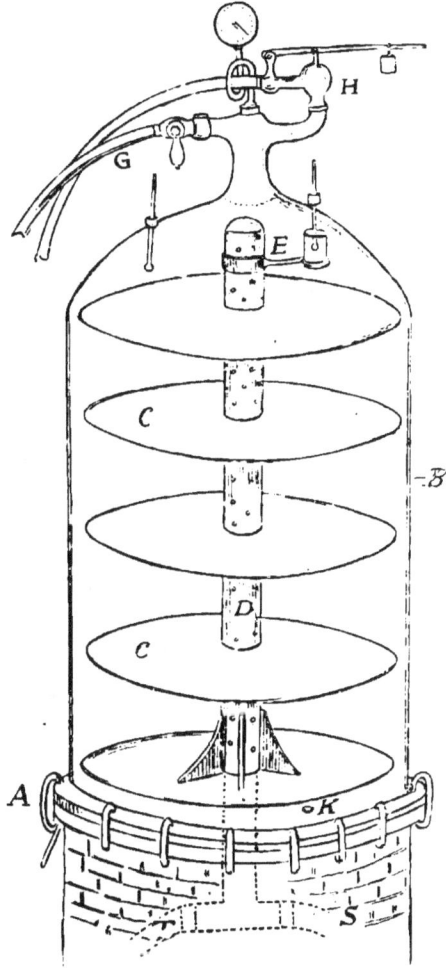

Fig. 18—BLACK FOREST STERILIZER.

of the operation, and is used again later when the required heat and pressure have been attained, so that

a continuous circulation of steam may be kept up in
the apparatus. A rubber tube is fastened to the end
of G and carried into a vessel with water to condense
the escaping steam. H is the safety valve. I, the
steam gauge. The bed plate is made concave, with
an outlet, K, to carry off the condensing water and
milk that may accummulate from breakage. The

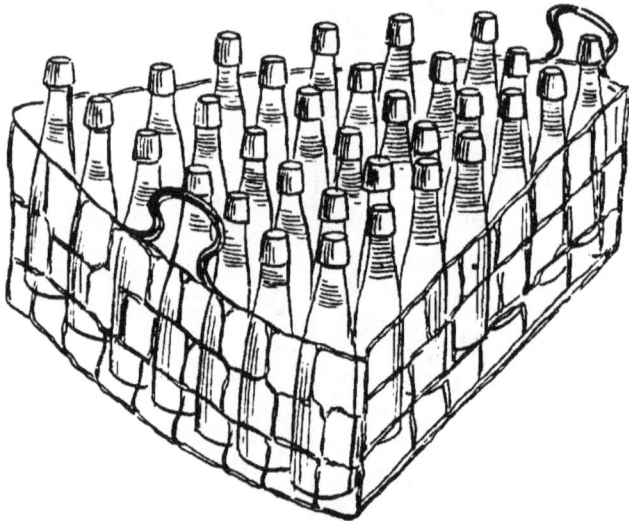

Fig. 19—MILK BOTTLES IN CARRIER READY FOR STERILIZING.

shelves are slightly convex for the same reason. The
bottles are placed in wire carriers, six of which fill
one of the shelves of the sterilizer. They are not
downright necessary, but will always be found a great
convenience and a saving in time and labor. A carrier
is shown in Fig. 19.

The duration of heating and cooling periods, which
together form one process of sterilization, are the following : One heating to 212° for thirty minutes, then

keeping for three hours at 95°, then heating to 212°
for another half hour, then cooling to 64° for ten
hours, then a final heating to 212° for forty-five minutes, and the cooling off to 58° as rapidly as the
bottles will stand. This rule for sterilizing should,
however, not be considered as fixed and unchangeable,
but it should be left to the investigation of the individual manufacturer of normal infants' milk to find

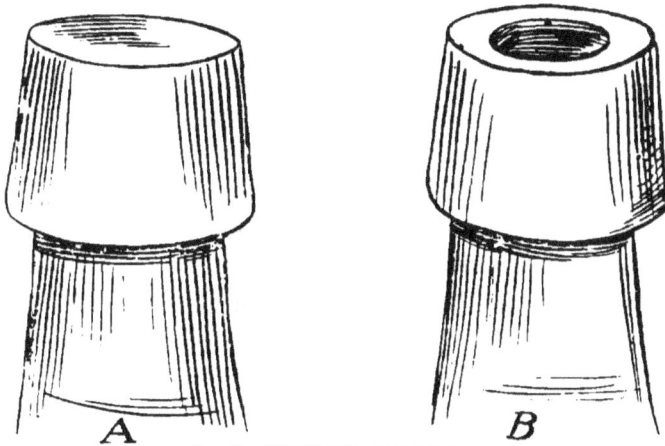

Fig. 20—AUTOMATIC SEALING CAP.

out, by trials, if the bacteria predominating in his
milk will allow of a modification or simplification of
the heating and cooling periods.

If the entrance of steam has been properly tem-
pered the breakage of bottles should be very small;
if, in spite of all care, there should result more than
one per cent. of breakage, then the glass is too brittle,
the bottles have been too rapidly cooled after manufacturing them. Before the second heating is com-
menced the hood is lifted and the bottles are inspected.

If the sealing by the rubber cap has been effective, this must be visible by the top of the cap showing a slight indenture. At times, when the heating has been too sudden, the violent escape of air from the bottles may have lifted the cap so that it does not show a concave; such rubber caps must now be pressed down again firmly and they will come out with hermetical sealing after the second heating.

The cooling must, every time, needs be accomplished very gradually, else considerable breakage will occur.

The last cooling should be to the lowest temperature attainable, a liberal supply of ice being an indispensible requirement of the establishment.

Immediately after withdrawing the bottles from the last heating in the sterilizer labels must be pasted on designating by their shape and color the grade of milk they contain.

RECAPITULATION OF MANUFACTURING PROCESS.

Cool the milk at once after drawing, to 40° F., unless there are milkers enough to keep the separator running from the start.

Test the fat percentage and acidity of milk.

Warm the milk to 86° F. previous to separating.

Separate and weigh cream and skimmed milk into one-third and two-thirds parts separately.

Calculate the quantities of cream and skim milk which have to be employed in the manufacture of grades I. and II., respectively.

Pour skim milk into the curdling vat and heat to 122°.

Place cream in cold water bath.

Add ferment to skim milk and let stand for fifteen minutes, then stir until curdling sets in, which should be about thirty minutes after time of adding the ferment.

Take out the paracasein at once.

Heat the remaining albuminous serum to 167°, and keep at this temperature for forty-five minutes, well covered.

Add the milk sugar, thoroughly stirring, then mix with the cream and sterilize.

For manufacturing the second grade, separate as for grade I., then divide skim milk as per calculation, add water, milk sugar and cream, mix thoroughly, bottle and sterilize.

Sterilize both grades equally. Keep in cool storage.

From every day's output of sterilized milk take two sample bottles, selecting one from the upper shelf of sterilizing apparatus and one from lower shelf, and place in bacteria incubator, properly labeled, for the purpose of ascertaining the keeping qualities of the milk ; and, also, if the sterilizer works equally well at top as it does at the bottom.

The greatest neatness and exactness should naturally prevail in executing all these operations, the manufacturer bearing in mind that he has guaranteed his product to be of a uniform standard of excellance, and that the normal infants' milk should show the same percentage of nourishing ingredients whenever it may be analyzed by a chemist.

ANALYSIS.

	Human Mlik. Per Cent.	Normal Milk, Grade I. Per Cent.	Normal Milk, Grade II. Per Cent.	Cow's Milk.
Fat . . .	3.2	3.0	3.0	3–5
Casein .	0.75	1.0	2.0	3.0
Albumen .	1.0	0.8	0.4	0.6
Milk sugar . .	6.25	6.0	5.7	4.8
Salt	0.4	0.6	0.5	0.7

To exclude all possibility of pollution by bacteria floating in the air of the laboratory (the mixing or the sterilizing localities) a disinfection of these premises should periodically be instituted. The safest and simplest way is by applying the fumes of formic aldehyd, a gas which kills all floating bacteria or germs. The lamp by which these fumes are generated is shown at Fig. 21. The vessel is filled with methyl

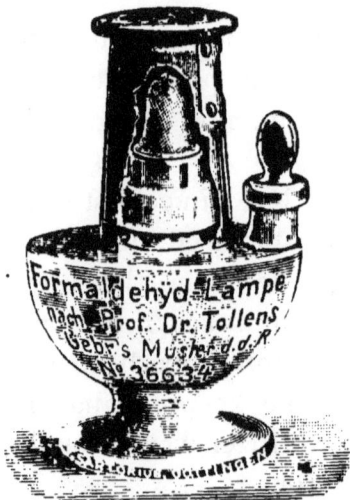

Fig. 21—DISINFECTING LAMP.

alcohol and the wick covered by a cap made of platina wire netting. After lighting the wick and waiting to see the platina netting become red hot, the flames is blown out when the glowing of the wire netting, however, continues producing a gas known as fumes of formic aldehyd. As soon as the fumes are strongly noticeable to our smelling organ, then the desired effect has been attained. The lamp is an invention of Professor Tollens, of Gœttingen, and may be procured through Messrs. Eimer & Amend, 205 Third avenue, New York city.

CHAPTER XI.

The Normal Dairy.

While no single part or ingredient of human food is of greater or equal importance and merits in its production in a higher degree strict supervision, yet none is consumed with a greater indifference as to its origin and pureness than cow's milk.

Considering the great advancements in the technical and scientific parts of dairying during the last decade, it is strange that the production of healthful infants' milk should have been so signally neglected. There exists no doubt to-day but what cow's milk is the best natural substitute for mother's milk and the best food for a child after weaning. Even if it were true that asses' milk would be preferable, there is too little of it ; or, if goat's milk were preferable on account of this animal's freedom from tuberculosis, yet the disagreeable taint peculiar to this milk, arising from the capronine it contains, makes it undesirable to most people, so that if there are other mammals whose milk, in its composition, comes closer to mother's milk, yet they are not of a kind either to furnish a sufficiency for our needs or they are not so domesticated as to allow us to draw it.

The conditions for the production of a healthy milk start with the selection of the cow, the feed she

receives, the degree of cleanliness she is kept in,
and in the treatment given at the hands of the dairyman.

As villages grew into towns and towns into cities
there would be found everywhere a class of people
that offered encouragement to the maintaining of
one or more dairies in close vicinity to the urban population. In many of the larger cities of the old continent dairy establishments had been maintained ever
since the beginning of the present century, and,
although they did not furnish anything else but
raw milk, such as was drawn from the cows, yet the
choice feeding and cleanliness practiced by these
dairies, which were under the daily inspection of the
patrons, insured a degree of confidence in the pureness of the product which allowed the dairyman to
charge such prices for his milk as would liberally re-
imburse him for the extra outlay encountered. Conditions allied to the mammoth growth of our modern
cities made it, however, impossible to increase the
number of these useful establishments, or even to
prolong the existence of the old ones. The high
value of building lots on one side, the hygienic ob-
jections to the accummulation of manure and the
difficulty to dispose of this valuable residue at a profit
on the other, have made these dairies disappear. The
control of quality of the milk that was then exercised
by the patrons now passed into the hands of the
health authorities and the police, and was extended
to all milk furnished for consumption, and it seemed

as if we had reached the boundary of the influence which we could exercise over the quality of marketable milk. We shall not here investigate what degree of efficiency this control has reached in general, or if it be sufficient to guarantee a fair quality for the milk of general consumption; as soon, however, as we come to the point to look at milk as a substitute for mother's milk, as a food for the new born-babe, we will from the perusal of the foregoing chapters agree that the present methods of control are of a glaring inefficiency.

It is, however, to be borne in mind that no change of method or added severity will be able to furnish the guarantee of pureness, which is so desirable, as long as milk has to pass through so many hands before it reaches the little consumer's mouth, and, that, at the time of its passing the milk inspector's test, it is only halfway, as it were, on the road which is strewn with possibilities of infection. If cow's milk is to be considered the only healthy substitute for the mother's breast, then our best efforts should be directed to produce this in the best form attainable. That no great success has been recorded, hitherto, in this direction may be largely attributed to the fact, that the difficulties to be overcome are located in so many different fields of work. Most farmers and dairy engineers lack entirely the necessary medical knowledge, and often, also, the support of the medical men, while the physician, if he manages to keep up with the complexity of tasks before him, is seldom in a position to

study the agricultural parts of the question or grapple
with the problems of technical dairying.

Every branch of production has, in its expanding
development, been forced to acknowledge the soundness of the principle of division of labor, yet if we
recapitulate what has been said about the necessary
supervision of the physical condition of the animals
furnishing the milk, about the necessity of sterilizing
it immediately after drawing, and about the pollution
it is exposed to by unclean handling before consumption, we will reach the conclusion that the production
of infants' milk is an exception to this rule of division of labor, and that no guarantee of pureness and
absolute healthfulness can be expected or given *unless*
the entire process of production, from the cow's
mouth to the baby's bottle, is covered by one and the
same responsibility, and controlled in every stage of
handling by those only competent to do so: the physicians and the veterinarian of the neighborhood.

We have seen that the purpose of sterilizing milk
is not only to give it keeping qualities by the deadening of all germs, also those of disease, but by this act
to make it healthy. The demand that sterilized milk
exclusively should be sold and used for the nourishment of infants and children is a just demand, because the delicate texture of the infant's intestines
more easily gives way before the irritations produced
by the bacteria and their exsudations. Besides, the
experiences of late years have forced upon us the
painful conviction that not infrequently there lurks
10

danger to health and life in the consumption of un-
sterilized or raw milk by the transfer of germs of disease. This experience is to be regretted so much
the more, as its recognition is connected with the fact
that this danger is inherent also to the progressive
development of our dairying industry, or at least,
that it is spread by it. There is no doubt but that
creameries, on the plan of association, are liable to
spread disease ; that they may be, and have been, the
medium to cause smaller epidemics, such as of typhus,
scarlet fever, etc., even though they possess all advantages of centralization and co-operation, they are,
however, not exempt from the great drawback which
adheres to all large institutions for distributing foodstuffs : the wholesale spreading and distributing of
disease.

But we need, most decidedly, protection against
such danger, and need it more particularly at such
times when the spreading of a disease has gained
larger dimensions, when the epidemic is rampant in
the houses of our cities and infection lurks behind
every imaginable vehicle. Ever since the study of
bacteriology has taught us that contageous diseases
are spread by bacteria or other low organisms, there
has been research on foot to investigate the roads on
which these infections move. Contrary to the former
belief that it was the local sanitary condition alone
that promoted a spreading, one has now cast suspicion
on the foods and beverages—water and milk—being

of universal consumption as the most likely promoters of infection.

But even, if in case of such emergencies, the local authorities should be able and competent to close such dairies or creameries to whose door the spreading of a disease has been brought home, this would not constitute a remedy, because the damage has already been done, as it is generally nimbler footed than the authorities. It is, therefore, to the preventive measures that we should turn our attention and efforts. More certainly is this true in regard to milk when we remember that it is apt to convey not only the germs of disease specific to mankind, but also some of those of the bovine species.

It would lead us too far from our subject if we should dwell on the methods that might be adopted for the prevention of infection by the means of milk, because, however urgently necessary they may be, still they might prove but too liable in their execution to seriously hamper and discourage an industry which it has taken the best efforts of the farmer, the scientist and the statesman to advance to the position of meritorious efficiency to which we have seen it lifted within the last few years.

Recognizing the difficulties that lay in the way of general disinfection of all milk brought to market we should turn to the next best expedient that offers : to produce and insure in the vicinity of every urban population, and within a distance of easy control, a certain quantity of milk especially reserved and treated for the consumption of infants. This idea has been partially carried out in a number of places where we hear of dairy farms furnishing "certified milk," an article purporting to be better and cleaner than other milk, and, as long as this certificate is one of real merit and not merely an advertisement, this milk is decidedly far superior to one of unknown origin, and its production a token of a very laudable spirit of enterprise—a step in the right direction— even if we know, from the foregoing, that such milk can lay no claim to being a healthy food for infants, inasmuch as it lacks being brought closer in its constituents to mothers' milk.

For the above named reasons the establishment of dairy farms for the production of prepared infants' milk, in close proximity to all urban populations, will, in the near future, receive greater attention, not only from the farmers, but, also, from the medical fraternity and the local authorities, from which parts they should receive all encouragement proportionate to the efficiency of their services.

The conditions to be exacted from such an establishment should bind the dairyman to the following stipulations :

1. To use no milk from any cow until eight days have elapsed after parturition ; nor from any cow six weeks before such event.

2. To use no milk from any cow in heat, off her feed, sick or any ways deranged, nor whilst being treated with strongly acting internal medicines.

3. To keep sick animals in a separate stable, tended by a special attendant.

4. To use the milk of any cow for no longer a period than seven months running.

5. To keep parturitant cows separated from the milking cows.

6. To keep neither horses, steers nor sheep in same stable with milking cows.

7. To feed milking cows on the most approved principles for avoiding acidity in milk, excluding all refuse feed, such as wet brewers' or distillers' grains or mash, adstringent oil cake or swill of any kind, and to water cows with pure water.

8. To feed to cows daily a proper allowance of salt.

9. To avoid all sudden changes in feeding, particularly from dry to green fodder and back, never to pasture milking cows but on artificial pasture of clovers and grasses, and to avoid all kind of feed or fodder having a laxative effect.

10. To keep cows scrupulously clean in comfortable, well ventilated stables, exercised, well bedded and kindly treated.

11. To exclude from the milk the first five strippings out of each teat at every milking.

12. To keep all milk free from any and all chemical admixture or adulteration, such as salt, borax, salicylic acid or others.

13. To keep no manure pile in close proximity of stables.

14. To enforce utmost cleanliness from all persons

engaged in milking, and handling milk, and to
enforce strictest abstinence from the use of tobacco
and liquor from all persons engaged in drawing, handling, preparing, or distributing milk.

15. To stop delivery of milk or collection of empty
vessels to and from all premises where infectious
disease is known to exist.

16. To superintend with untiring vigilance the
cleansing and sterilizing by steam, hot water and soda
of all utensils and apparatus used in handling, preparing and conveying milk.

17. To engage the services of a competent veterinarian for the frequent inspection and investigation
of the sanitary condition of the milk cows, and furnish clean bill of health every month from the veterinarian for all cows whose milk is used in preparing
the normal infants' milk.

18. To facilitate in every way, in all premises and
at all times, the thorough inspection of the entire establishment by members of a committee of the medical profession, or the local board of health.

It will be conceded that the proposed conditions for
the production of pure milk can easily be fulfilled
without incurring great expense, and this is a requirement that should not be lost sight of, for, in fixing
these stipulations, a reasonable limit to precautionary
measures must be admitted, without which, the consequent considerable increase in cost of production
would tell on the price of the milk, tend to put it
beyond the reach of the poorer classes, and thus frus-

trate to a considerable degree the good for which the establishment had been created. It is well to remember that conditions which might appear ideal to the medical mind may be absolutely impracticable of execution.

However plain the detrimental effects of common impure milk may be to the life in general, and to that of infants in particular, the entire bearing of the matter and the importance of ameliorating such conditions is not recognized by the masses of the population, nor will the public be found willing to pay a higher price for infants' milk as long as the entire *visible* amelioration would consist in a new-fangled stopper on the bottle or in a colored label around its neck.

The subtelty and the minuteness of the noxious germs contained in ordinary cow's milk, and the impossibility of furnishing a daily certificate of their deadening or removal, based on the finding of a chemical and a microscopical investigation, make this business, in a great degree, one of confidence placed by the public in the honesty of the dairyman. But experience has shown that even the greatest honesty on the part of the dairyman and his skill in sterilizing is not in all cases sufficient to insure an untainted milk to an infant, because all precautions are futile if the sterilized milk, prior to its consumption, is left to the manipulation of careless and unreliable persons.

This is one of the reasons why infants' milk should be furnished in hermetically closed small bottles of a shape to allow the adjusting of the feeding nipple immediately after removing the stopper shortly before warming and using the milk. Although small sterilizing apparatus exist, and may be bought, yet, for reasons previously demonstrated, they can by no means be considered as giving the same security of a dairying and sterilizing establishment, and German scientists agree that the manufacture of infants' milk cannot be conducted with any degree of success in the household of the consumer, or by parties not perfectly versed in the functions or properties of the different ingredients and equipped with the most perfect appliances that will insure the production of an article of uniform composition and merit.

Other reasons pointing toward the advisability of entrusting a larger establishment with the manufacture of infants' milk are that—

1. By the use of the cream separator a large percentage of the most noxious germs are retained in the bowl of the machine, imbedded in the separator slime.

2. The percentage of fat contained in the fresh milk, to be converted into infants' milk, can be ascertained and regulated daily before and after manufacturing the infants' milk.

3. All mixtures are performed with greater accurateness and precision, because everything is done by exact weight and measure, and not by table or teaspoonfuls.

4. All mixing, sterilizing and cleansing is done more efficiently, quicker and cheaper.

5. All materials used are procured wholesale, at a considerable reduction in price, which tells on the price of the milk.

After reviewing the points which could make such an establishment, or a number of them, a desirable acquisition to the neighborhood of an urban population, it is but fair to ascertain if this will, under existing circumstances and conditions, equally be a desirable undertaking for a dairy farmer. Binding himself to the afore enumerated clauses, for the conduction of his establishment, he is certainly entitled to the moral and efficient support of the authorities and the board of health. The guarantee of pureness, which is given to the products of the establishment by a constant or periodical supervision, is absolutely necessary to guard the public from imposition, as well as the dairyman from the appearance of a spurious article, which would at once tend to destroy his undertaking by discrediting normal infants' milk through the rapacity of unscrupulous rival parties. For the same reason, the retailing of normal infants' milk should not go through the channel of the small milk trade, but through the establishment itself, through a designated number of drug stores or large milk traders. This business is one of confidence, because of the difficulty of daily testing the pureness of its products, it is, therefore, natural that it be undertaken by, or conceded to, only such parties who— apart from their physical and financial ability to personally superintend and foster it—have thoroughly mastered the theoretical and technical parts of the matter and can command the entire confidence of the " parties of the second part." On the other hand, it would be folly for a dairyman to undertake the fitting out of a sterilizing establishment without the encouragement and support just mentioned ; it seems, however, unnecessary to dwell longer on this subject; wherever undertaken, by the proper person and with the proper appliances, the advantages that may accrue to the sanitary condition and the welfare of the population it would serve, have been spontaneously recognized. As an instance I will mention that it is a well established fact that since the establishment of the dairy of Mr. Bolle, in the German capital, the morality of the infants has been lowered twenty-five per cent.

As to general rules for the location of such an establishment, they will, in a great measure, always be govered by local conditions, it should, however, cerly not be located at a greater distance from the population which consumes its products, than will allow of an easy supervision and rapid transportation. This distance will be regulated, in a manner, by the value of land in the vicinity of the city or town it would have to serve. The advantages which close proximity may confer are entirely lost if the price of the milk has to be raised to meet the extra expense of high rents on land, and as long as transportation can be expeditously carried on, there need exist no other limit to the distance but that set by the possibility of effective medical control of the establishment.

As regards transportion, it is well to remember that bottles with normal milk must never be filled to the brim, as part of the milk would boil out during sterilization; they will, therefore, not stand protracted shaking on rough roads as raw milk would, because the butter fat easily collects in the neck of the bottle and butters out.

In the time of old town dairies, a considerable in-

Fig. 22 –SIMMONTHAL SWISS BULL.

fluence was accorded to the breed of cattle which should be kept by such furnishing milk for infants; on the old continent, England excepted, it was generally believed that the Alpine breeds were the healthiest, and, therefore, the only proper breeds to furnish such milk; since we have, however, learned to covert the milk of any healthy cow into a milk, which, in all its nourishing constituants, is identical to the human milk, irrespective of the relative proportions contained thereof in cow's milk, this question of breeds has lost a great deal of its importance, the main requsite now being : a healthy cow.

The relation of fat to casein and of total percentage of solids to that of albumen is, however, a variable quantity in the different breeds, and should be studied and taken into account when planning the manufacture of normal infants' milk. The work of a number of experiment stations on this line has been invaluable in determining the respective percentages in the milk of the standard breeds of cattle.

The average composition found by analyses of 28,000 samples of milk was total solids, 12.68 per cent.; fat, 3.91 ; solids, not fat, 8.77 ; specific gravity, 1.0318. When computed for an entire period of lactation, the following figures were found for the respective breeds :

BREED.	Number of Analyses.	Water, Per Cent.	Total Solids, Per Cent.	Solids, not Fat, Pr. Ct.	Fat, Per Cent.	Casein, Per Cent.	Milk sugar, Per Cent.	Ash, Per Cent.	Nitrogen, Per Cent.	Daily milk yield, lbs.
Holstein-Friesian	132	87.62	12.39	9.07	3.46	2.39	4.84	0.735	0.540	22.65
Ayrshire	252	86.95	13.06	9.35	3.57	3.43	5.33	0.698	0.543	18.40
Jersey	288	84.60	15.40	9.80	5.61	3.91	5.15	0.743	0.618	14.07
American Holderness	124	87.37	12.63	9.08	3.55	3.39	5.01	0.698	0.535	13.40
Guernsey	112	85.39	14.60	9.47	5.12	3.61	5.11	0.733	0.570	16.00
Devon	72	86.26	13.77	9.60	4.15	3.76	5.07	0.760	0.505	12.65
Average		86.37	13.64	9.40	4.24	3.58	5.09	0.731	0.534	16.20

"According to the above table the ash varies least among the above constituents of milk, sugar next, then casein, and fat by far in excess of all, varying over four times as much as casein."

The average composition of the total solids was as follows :

Average Composition of Total Solids of Milk.

BREEDS.	Total Solids, Per Cent.	Solids, Not Fat, Per Cent.	Fat, Per Cent.	Casein, Per Cent.	Sugar, Per Cent.	Ash, Per Cent.
Holstein-Friesian.....	100	73.2	28.0	27.4	39.1	5.93
Ayrshire.............	100	71.6	27.3	26.3	40.8	5.34
Jersey...............	100	63.6	36.4	25.4	33.4	4.82
American Holderness	100	71.9	28.1	26.8	39.7	5.53
Guernsey............	100	64.9	35.1	24.7	35.0	5.16
Devon...............	100	69.7	30.1	27.3	36.8	5.52
Average..........	100	69.2	30.8	26.3	37.5	5.38

The variation in the percentage of fat in the total solids is larger than for any other constituent.

The following tables give the results of investigations by the New York Experiment Station for the production of milk only, as the results for the separate breeds materially differ when it comes to the production of cream, butter and cheese.

Tabulated Summary Showing Relative Results of Comparison for Different Breeds of Cattle with Reference to Production of Milk. Figures based on Lowest results as 100.

	American Holderness	Ayrshire	Devonshire	Guernsey	Holstein Friesian	Jersey	Shorthorn
Relative cost of food eaten	114	131	100	123	135	121	123
Relative amount of milk produced	144	172	100	135	199	127	152
Relative cost of milk	117	114	145	232	100	139	120
Relative amount of milk solids produced	125	151	100	139	162	134	150
Relation of per cent. of milk solids	107	108	123	126	100	130	121
Relative cost of milk solids	111	106	122	107	102	110	100
Relative value of milk at 1.28 cents per lb	144	171	100	135	199	127	142
Relative value of milk based on solids at 9½ per lb	125	151	100	139	162	134	150
Relative value of milk based on fat at 26⅓ cents per lb	116	134	100	156	145	154	149
Relative apparent profit from milk	151	194	100	177	224	150	211
Relative actual profit from milk	163	214	100	202	255	171	245

	American Holderness	Ayrshire	Devonshire	Guernsey	Holstein Friesian	Jersey	Shorthorn
Number of cows	2	4	3	4	4	4	1
Total number of periods of lactation	4	12	5	6	4	11	2
Cost of food eaten	$42.90	$49.32	$37.52	$46.15	$50.73	$45.49	$46.22
Pounds of milk given	5721	6824	3984	5385	7918	5045	6055
Cost of milk in cents per pound	0.76	0.74	0.94	0.86	0.65	0.90	0.78
Cost of milk in cents per quart	1.63	1.58	2.02	1.85	1.39	1.95	1.68
Pounds of milk solids produced	724.1	860.4	577.4	804.0	936.5	755.4	866.2
Per cent. of solids in milk	12.66	12.74	14.50	14.93	11.88	15.37	14.30
Cost of milk solids in cents per pound	5.93	5.68	6.50	5.73	5.42	5.87	5.34
Money value of milk at $1.28 per pound	$73.22	$87.34	$51.00	$68.93	$101.35	$64.58	$72.50
Money value of milk based on milk solids at 9⅓ cents per pound	67.58	81.14	53.89	75.04	87.41	72.37	80.85
Money value of milk based on milk fat at 26⅔ cents per pound	56.12	64.47	48.27	75.18	70.07	74.30	72.03
Apparent profit (money value of milk less cost of food)	24.69	31.73	16.37	28.88	36.65	24.63	34.60
Calulated value of skim-milk	15.61	19.06	12.00	15.81	20.49	13.78	18.20
Market value of skim-milk	7.81	9.53	6.00	7.90	10.25	6.89	9.10
Actual profit (apparent profit less market value of skim-milk)	16.89	22.20	10.37	20.97	26.40	16.74	25.50

When we turn to the question as to which breed of cows will be the most economical for the production of the normal infants' milk, we must bear in mind that the constituants of the milk we should produce are fixed quantities, and that no considerations of preference for any particular breed should interfere in the decision.

Considerable controversy has also arisen over the physical condition of the cow, in respects to her ability to produce a pure milk, unimpaired by such changes as arise from the collateral functions of the generative organs, the strictest doctrinarians advocating the exclusion of all animals in a state of pregnancy, and this exaction has been and can be fulfilled by dairy farmers situated in localities where cows may be advantageously disposed of to the butcher after finishing their period of lactation, but this condition does, more generally, not prevail in the neighborhood of those populations that stand in the most urgent need of a normal dairy establishment and, where the exactment of such a stipulation would mean a loss of, perhaps, fifty per cent. on the value of the cows and, correspondingly, demand the reimbursement of this loss by an advance on the selling price of the milk.

As to feeding the cows, it should be made the rule to feed only morning and evening and to avoid feeding dry roughage during the time of milking.

Although the size and manner of construction of the stable, or barn, in which the cows are kept is not of a direct influence on the quality of the milk produced as long as it is well arranged, properly lighted and ventilated, yet there are some reflections of importance which should be considered in connection therewith. In the columns of our agricultural and dairying periodicals we frequently come across the discription of so called "model barns," the model part of which varies, however, as to the point of view from which the owner has started in erecting it. Many of them consult only their own advantage, others try to make their cattle comfortable, some try to combine the interest of both owner and cattle, very few, however, pay any regard to the interest the consuming public may have in the construction of the barn. A barn may be admirably planned for economical management; when the cattle are, however, fastened in stanchions on cramped platforms their welfare has not entered on the "model" arrangement, or if a barn, with an otherwise faultless arrangement, stores the manure in a cellar beneath it, then the interest of the public has not been taken into account in laying out the model part of this barn, because it makes it unfit to produce pure and untainted milk, such as we should insist on for the production of normal infants' milk.

When a farmer or dairyman has no other interests to consult but his own, when building a new barn, he is free to indulge in any eccentricities that may be prompted by a variety of motives, some based on practical experience and economical calculations, others again, however, on motives far less meriting of imitation. I always feel a genuine pity for the possessor of a very large barn, a few of which I have seen, and seen photographs and descriptions of many more, particularly located in this country; they are, in most cases, very creditable testimonials to the designing carpenter's skill, and pretty board and shingle monuments to the owner's length of purse, but as for their usefulness and merit for an establishment producing infants' milk after the methods herein described and under the supervision of or under contract with a medical board, they should be entirely condemned. The normal dairy must not only be able to supply the requisite infants' milk, it should also be regulated in a manner to offer the greatest possible security for maintaining this supply continuously, because a sudden falling off from it might mean interrupted development and serious inconvenience to many, and, perhaps, death to some infants. This security is not found in the large barns or stables, where a disaster may sweep off the entire productive force in a few hours, or where an infectious disease brought in by one animal may—while in its latent period and, therefore, undetected—spread and infect every animal in the whole herd. Therefore, when there is a chance to do so, it is advisable to keep the cows in separate barns, none to exceed thirty head. Newly bought animals, if not coming from stables in close proximity to the farm and from herds notoriously free from all disease, should be kept confined

separately for a term of ten days. Whoever has had
a chance to experience the trouble which epidemic
abortion gives, its pugnacity and infectious character,
will never advocate the building of a mammoth barn.
Besides which, the limited number of cows mentioned
above is just the number to be well cared for by one
man, and I have ever found that attendants will work
better and give more care when they know that the
responsibility for any neglect cannot be loaded onto
" the other fellow." A good man will be proud of
the good looks and thrift of his animals, because he
knows that the credit for it is earned by himself alone.
All over the Old Continent the Swiss are renowned as
being the best milkers and attendants on cattle.
From my own experience, and from the testimony of
hundreds that employ them, it is a well merited re-
nown, so much so that in several countries any attendant on milk cows is termed a " Swiss."

Finally, the question may arise how is the dairyman, who intends taking in hand this branch of business, to insure himself and his undertaking in these
times of hand to hand fight in competition against
the multitude of those who, though too indolent or
too careful to risk any capital in a new and untried
industry at the start, yet fall upon it as on a legiti-
mate prey as soon as they see their neighbor making
a success of it. Unrestrained competition will, in all
instances, tend to lower the standard of efficiency
and merit in any product of general consumption,
the quality of which cannot be judged by the outer

appearance. If the advantages to be gained by an urban population from the establishment of a normal dairy are not recognized as meriting protection and support, then the dairyman is located near the wrong place. Not a single instance has, however, come to my knowledge of this ever happening. Quite the contrary; these establishments have, particularly in Germany, multiplied rapidly, owing to the hearty and effective support received at the hands of the medical fraternity.

CHAPTER XII.

Conclusion.

However advantageous and promising an undertaking may appear, yet exhaustive investigation and calculations of cost of production, and probable amount of sales, should form a principal factor in the decision. The dairyman intending to take up this industry, should first of all find out if the physicians of the place take an active interest in the matter. This is generally the case, as no doctor can afford to ignore or treat the subject with indifference; moreover, infants are, in most cases, the most ungrateful patients they have. The next step is to find out the number of residents who would, in all probability, be found willing to pay a higher price for a healthy infants' milk. On an average we may calculate on forty births a year for every 1,000 inhabitants. We may further calculate that ten of these new-born infants will be nourished with normal milk for the entire first year, and twenty for a period of six months only. In the second year of their lives, infants should be able to take pure cow's milk, this should, however, always have been produced under observation of all precautionary measures mentioned heretofore, and always be sterilized. Let us calculate that for twenty children, in their second year, such sterilized cow's milk would be demanded, we would then figure on a total daily demand per thousand inhabitants, as follows :

10 Infants in their 1st year, at 0.75 qts. 7.5 qts.
10   "   "   "   1st   "   " 1.00 "   10.0 "
20 Child'n "   "   2d   "   " 1.00 "   20.0 "
                                           _____
                                           37.5 qts.

This would be the milk necessary for infants in their first and second years, in many places, however, the consumption of normal infants' milk, and sterilized cow's milk, has risen to fifty quarts per 1,000 inhabitants daily, owing to a demand, for dyspeptics, and older children. From these quantities we may judge that, even in smaller places, the establishment of the manufacture of normal milk may be remunerative, particularly as it may be sent to adjoining places without spoiling. Experience has shown that in all cases there has been a steady increase in the demand. To encourage the introduction, medical men must be furnished with the means of testing the normal milk in their practice. Printed matter, setting forth the merits of the normal milk, should be mailed to all families where an infant has been born, and an arrangement can generally be made to receive the address of such families from the office of registration.

In many instances the furnishing of normal milk to poor mothers, is a favorite way of bestowing charity, and checks should be printed for the receipt of stated quantities of milk, to facilitate this, and to avoid the giving of cash, which is apt to be preverted to other uses. It will be found convenient to deliver the bottles in light wooden boxes, holding from fifteen to twenty-five bottles each, the number varying with the size of the bottles.

Fig. 25—CLEANSING BRUSH.

Some trouble is experienced at the beginning with the returning of the bottles and rubber caps, and some strictness is required, on the part of the dairyman, to oblige the patrons to return the bottles clean, or what this may mean to the consumer. We know that real cleansing means the application of steam, hot water, soda and the brush. This is a point of the greatest importance. The return of clean bottles must be insisted upon at all hazards. In connection with this, and to illustrate the baneful effects of unrestricted competition, I will mention my experience when walking along Fifth avenue, New York City, in May of this year. From a milk wagon, gorgeously appointed, a clean man was distributing dainty glass jars with milk to the basements of different residences; it struck me as a model arrangement, until I saw the man return with a load of empty jars. They had not been cleaned after emptying out the milk, and were in a state of disgusting filth and sourness. I imagine that if this milkman would object to receiving the bottles in this disgraceful condition the family would speedily find another milkman, less fanciful.

Fig. 26—RINSING VAT.

As for the premises required by the establishment, they should be of the same size as a creamery handling the same quantity of milk. There should certainly be four separate rooms, the first for the receiving vat, cooler, heater and separator; the second for the mixing, weighing and bottling; the third for the sterilizer; the fourth for the cleansing of bottles and utensils. All floors should be cement laid, and on the same level, so that trucks carrying milk or bottles

may be wheeled from one room to the other without
obstruction.   Ice house and storage should be close
by.

The cost of putting up and fitting an establishment
of this kind can hardly be closely estimated for general direction, as they will change for every locality ;
the principal items of expense may, however, figure
under the following :

Steam boiler	$300 00
Babcock fat tester	15 00
Milk heater	45 00
Milk cooler	45 00
Cream separator.	225 00
Two bottle cleaning machines	28 00
Filling apparatus	40 00
Sterilizer	300 00
Bacteria incubator.	36 00
Table and platform scales	50 00
Bottles and rubber caps	250 00
Thermometer and other glass instruments. .	24 00
Mixing vats	80 00
Smaller utsenils	35 00
Packing cases, labels, printing, advertising .	100 00
Steam and water pipe brass, work	125 00

There is no absolute necessity for a steam engine,
because the cream separator, which is the only machine used requiring power, can be bought with steam
turbine, an arrangement which, for our purposes,
must be recommended.

The price which the dairyman is to receive for normal milk will be regulated, in some degree, by the

price which common good cow's milk is obtaining at
retail, and by the average amount of prosperity of the
place. In a majority of cases the normal milk may
be manufactured and sold at an advance of from fifty
to seventy-five per cent. on the retail price of cow's
milk, although, in many instances, double the price
of ordinary milk is obtained. It seems needless to
dwell on the necessity of a liberal supply of water for
the uses of the normal dairy, the cleaning of the bottles alone requiring a considerable quantity. Where
cool spring water cannot be counted upon all the
year round, ice must be brought into requisition.
This will always be a necessity in warmer climates,
and it is just in these that the amelioration of existing conditions for the production of a healthy infants'
milk is the most urgent.

Fig. 27—COMBINED BRUSH AND RINSER.

Short courses of practical instruction will be or-
ganized, as purely theoretical instruction has proved

inadequate to impart that degree of security which is an indispensible condition to success for everyone contemplating the manufacture of normal infants' milk.

There can exist but little doubt that the near future will bring into greater prominence the agitation now so ably sustained by a number of scientists, who, working on this field of investigation, are the truest benefactors to infant mankind.

The enactment of stricter codes for milk inspection, the rigid enforcement of those already existing, the tuberculin test for all milk cattle, the pasteurization or sterilization of all merchantable milk, and the manufacture of artificial mothers' milk, will soon be demands in universal requisition ; it will be for the enterprising and intelligent dairyman to watch his chances, to keep abreast of the times he is living in, by considering whether existing circumstances do not warrant his embarking in this manufacture. Here is the chance, so seldom offered in our profession, for a man to lift himself above the great horde of competitors, by intelligence and progressive energy in producing an article, the success of which will depend on the theoretical and practical training of his mind and business capacity, more than on his aptitude to hold a plow, handle a pitchfork, or follow in the footprints of his forefathers.

COMPARISON—WEIGHTS, MEASURES AND THER-
MOMETERS.

One American gallon is equal to 4 quarts (a 2 pints.

One American gallon is equal to 8 pints (a 16 ounces.

One American gallon is equal to 128 ounces (a 8 drachms.

One barrel holds 31½ gallons.

One hogshead holds 63 gallons.

One tierce holds 42 gallons.

One puncheon holds 84 gallons.

One gallon is equal to 1,453 liter.

One gallon is equal to 3,785 cub. centimeter.

One gallon is equal to 10 pounds of water.

One Engl. Imp. gallon contains 277 cub. inches.

One ale gallon contains 282 cub. inches.

One wine gallon contains 231 cub. inches.

One dry gallon contains 268 8-10 cub. inches.

One bushel has 2,150 4-10 cub. inches.

One quart dry measure is equal to 2½ pounds milk.

One quart dry measure is equal to 1 1-7 quart liquid measure.

One normal quart weighs 2.15 pounds.

100 pounds of milk is equal to 47 quarts.

One pound Troy is equal to 12 ounces, each 8 drachms, each 3 scruples, each 20 grains.

Fahrenheit.	Reumur.	Celsius.
$+257.0$	$+100.0$	$+125.0$
248.0	96.0	120.0
230.0	88.0	110.0
212.0	80.0	100.0
194.0	72.0	90.0
176.0	64.0	80.0
158.0	56.0	70.0
140.0	48.0	60.0
122.0	40.0	50.0
104.0	32.0	40.0
86.0	24.0	30.0
68.0	16.0	20.0
50.0	8.0	10.0
32.0	0.0	0.0
$+14.0$	$= 8.0$	$=10.0$
$= 4.0$	$=16.0$	$=20.0$
$=22.0$	$=24.0$	$=30.0$